Becoming Neolithic

Becoming Neolithic examines the revolutionary transformation of human life that was taking place around 12,000 years ago in parts of southwest Asia. Hunter-gatherer communities were building the first permanent settlements, creating public monuments and symbolic imagery, and beginning to cultivate crops and manage animals.

These communities changed the tempo of cultural, social, technological and economic innovation. Trevor Watkins sets the story of becoming Neolithic in the context of contemporary cultural evolutionary theory. There have been 70 years of international inter-disciplinary research in the field and in the laboratory. Stage by stage, he unfolds an up-to-date understanding of the archaeology, the environmental and climatic evidence and the research on the slow domestication of plants and animals. Turning to the latest theoretical work on cultural evolution and cultural niche construction, he shows why the transformation accomplished in the Neolithic began to accelerate the scale and tempo of human history. Everything that followed the Neolithic, up to our own times, has happened in a different way from the tens of thousands of years of human evolution that preceded it.

This well-documented account offers a useful synthesis for students of prehistoric archaeology and anyone with an interest in our prehistoric roots. This new narrative of the first rapid transformation in human evolution is also informative to those interested in cultural evolutionary theory.

Trevor Watkins is an emeritus professor at the University of Edinburgh where he taught prehistoric archaeology of the east Mediterranean and southwest Asia for many years. He led excavations in Cyprus, Syria, Iraq and Turkey. For many years, his research has been focused on the Neolithic period in southwest Asia, between twelve and eight thousand years ago. In recent years, he has been concentrating on relating the results of recent archaeological research to the latest theoretical work by leading researchers in cognitive archaeology and the field of cultural evolutionary theory. If evolution takes place over time, then archaeologists should be able to document and calibrate the process with material evidence.

Becoming Neolithic

The Pivot of Human History

Trevor Watkins

LONDON AND NEW YORK

Designed cover image: Courtesy of the Aşıklı Höyük Research Project

First published 2024
by Routledge
4 Park Square, Milton Park, Abingdon, Oxon OX14 4RN

and by Routledge
605 Third Avenue, New York, NY 10158

Routledge is an imprint of the Taylor & Francis Group, an informa business

British Library Cataloguing-in-Publication Data
A catalogue record for this book is available from the British Library

ISBN: 978-0-415-22151-1 (hbk)
ISBN: 978-0-415-22152-8 (pbk)
ISBN: 978-1-351-06928-1 (ebk)

DOI: 10.4324/9781351069281

Typeset in Times New Roman
by codeMantra

Contents

Figures

Acknowledgements

As I finally – finally – complete the writing of this book, I am deeply aware of the help, support, and encouragement that I have had over the years that I have struggled with it. The incentive was planted when I was invited by the Society of Antiquaries of Scotland to give the six-lecture series of Rhind Lectures in Edinburgh in 2009. Out of that came the invitation to submit a paper summarising the lectures to the journal *Antiquity* (Watkins 2010). There is a long tradition that encourages a senior scholar to turn their Rhind Lectures into a book that records his or her mature understanding of their subject. It has taken me too long, but I express my gratitude to the Society for the encouragement that the Rhind Lectures invitation gave me and apologise for my dilatory response.

Initially, I was fortunate in forming a friendship with Klaus Schmidt, the excavator of Göbekli Tepe until his sudden death in 2014. We worked together on the idea of a research project that sought to understand what drove the profound cultural and social changes that marked the beginning of the Neolithic period, producing monuments and complex symbolism such as that of Göbekli Tepe and other sites in the region. For both of us, the project was the challenge to follow up on the ideas of Jacques Cauvin (1994, 2000) and his provocative hypothesis of 'the birth of the gods' in a 'psycho-cultural' revolution at the core of the new world of the Neolithic. Our research proposal was welcomed by the John Templeton Foundation. Part of their funding enabled the Göbekli Tepe excavation project to carry out a radiocarbon dating programme that demonstrated categorically that the monumental structures and the extraordinary sculptured symbolism indeed belonged in the early Pre-Pottery Neolithic. Another part of the project was a workshop held in Urfa, the city in Turkey close to the site of Göbekli Tepe. That enabled a group of Neolithic specialists to meet and talk with several scientists from other disciplines and to spend a day at the site of Göbekli Tepe in the company of Klaus Schmidt. That workshop certainly excited the thinking of a number of us, myself especially. We also planned a multi-authored book about the Epipalaeolithic–Neolithic period, to be launched through the medium of a workshop in Berlin. In the wake of Klaus Schmidt's sudden death, we did manage to hold the workshop; but, in the difficult situation of the Göbekli Tepe project for the German Archaeological Institute consequent on his death, the plans for the book fell apart. I recall that our proposal to the Templeton Foundation included our hope that through our project we could 'devise new inter-disciplinary modes in prehistoric research, and seek to change the way that archaeologists and the public understand an important formative period in human history'. I hope that this book will, to some extent, stand in for the big book that was at the core of our project and help to advance how we recognise the importance of the Epipalaeolithic–Neolithic transformation in human history. I am very happy to acknowledge my debt to the John Templeton Foundation, whose staff continued to help and encourage me.

I am equally happy to acknowledge my debt of gratitude to many of my archaeological friends who have helped and encouraged me over recent years that have been difficult for me personally. I am especially grateful to all those colleagues who have sent me images from their field research archives to be used as illustrations; their names can be found in the credits in the picture captions. And finally, I wish to record my deep gratitude for the love and support of my family through hard times of recent years. Without their positive encouragement, this book could easily have been stillborn. And I dedicate the book to my wife Antoinette, with deep regrets that I was not able to complete it for her before she died.

Introduction

The Neolithic in southwest Asia (or the Neolithic revolution in the Near East, as it used to be called) has been a major focus of research for more than 70 years. The idea that the Neolithic, or agricultural, revolution was a pivotal moment in the whole of human history, comparable in importance to the industrial revolution of the nineteenth century, has been 'general knowledge' for many people over generations. However, throughout all this long time, since Gordon Childe (1936, 1942) proposed the idea, there have been questions, debates, and arguments about the nature of this so-called Neolithic revolution. These continuing arguments suggest that, despite the decades of research, a satisfactory conclusion had not been reached. Is that because we still need more information to answer our questions? Why did arguments continue about what really happened to make this revolution? Why did it happen then (roughly, at the beginning of the Holocene era, starting 12,000 years ago), and why not earlier?

This book has been gestating over decades, but it has been completed at a critical moment, when four trends have converged to make a new solution to these questions possible. Firstly, because of all the exploration and excavation that has been going on, of course we have so much more information, and fewer blanks on the map of exploration. Secondly, because the researchers working on all this new data have applied new and increasingly sophisticated analytical tools, we have been rewarded with new and often unexpected insights. Thirdly, because academic debate among archaeologists has refined our understanding: some theories have failed to stand up to criticism. Fourthly, and most importantly, the advances in our understanding of the Neolithic have converged with recent advances in evolutionary theory. In the last 20 years, cultural evolutionary theory has blossomed and grown into an exciting new sub-field that offers a quite different take on the story of human evolution. I have been following the developments in cultural evolutionary theory since the 1990s and attempting to write about new ways to integrate these developing ideas into the archaeology of the Neolithic. The last chapters here show how cultural evolutionary theory and cultural niche construction theory enable us to make sense of the Neolithic as the critical turning point in the whole of human history.

There are two matters that it would be good to deal with before we go on into the core of the book. The first concerns the necessary technical terms, like 'Neolithic',

DOI: 10.4324/9781351069281-1

that I have to use, and the commonly used terms that I avoid. The second is an outline of the structure of the book, so that the reader may know where they are and what lies ahead.

The archaeological period known as the Neolithic, both in Europe and what used to be called 'the Near East', has long been identified with the name of Gordon Childe (who in 1927 founded the Archaeology Department at the University of Edinburgh, where I spent most of my career). Childe produced academic books that became the standard texts for students for 30 years. But he also published popular paperbacks that established the fundamentals of his Neolithic and urban revolutions in the public mind (Childe 1936, 1942). The Neolithic revolution, he said, was a prehistoric transformation that saw the emergence and spread of the first village-societies and the first farmers. Before Childe, the Neolithic, the New Stone Age, was defined by new kinds of stone tools, and the appearance of new craft skills, such as pottery, the plough, and the weaving of textiles. For Childe, the Neolithic revolution was a social and economic transformation equivalent in its historical importance to the industrial revolution, which had so impressed Friedrich Engels and Karl Marx.

The Neolithic revolution has frequently been referred to as 'the origins of agriculture' (to name but a few: Bar-Yosef & Meadow 1995; Peake 1928; Rindos 1984) or 'the agricultural revolution' (e.g. Barker 2006). It is true that there has been a great deal of productive research on the domestication of plants and animals, but I take exception to the reduction of this major transformation to the switch from foraging to farming. I reject that approach for three reasons. Firstly, farming, as we commonly understand the term, did not begin in the Neolithic. As will become clear in Chapters 2 and 3, while people began to cultivate crops even before the Neolithic began, and to manage flocks of sheep and goat in the Neolithic period, farming as we usually think of it – with fields that were ploughed with ox-drawn ploughs, producing harvests that were brought home in carts or wagons – began after the end of the Neolithic of southwest Asia. What developed within the Neolithic period itself was effectively horticulture, or 'garden agriculture', where people began to work the land by hand to produce enough for their needs. Secondly, to reduce the transformation to a change in the subsistence economy is like explaining the industrial revolution as the replacement of water wheels by the steam engine. There is so much more to the transformation process than simply the change from hunting and gathering to agriculture and the keeping of animals.

The transformation that is our subject here happened in southwest Asia, and in five or six other parts of the world, according to Bruce Smith (1994), or as many as ten different regions according to Fuller (2010). Price and Bar-Yosef (2011) introduced a supplement to *Current Anthropology* with a battery of papers on 'The Origins of Agriculture: New Data, New Ideas' worldwide. The southwest Asian transformation, beginning ten millennia before the end of the Pleistocene period, was the earliest, and it has been under the most intense investigation over the longest time. It has been the field of research where technical and scientific advances have been pioneered, and it has become the paradigm for the process.

I deliberately avoid Childe's term 'Neolithic revolution' and the word 'revolution', which some people misunderstand as implying that it happened suddenly, and other people find suspect and politically provocative, largely because of Gordon Childe's engagement with Marxist historical materialism (and socialist politics). I also prefer the term 'southwest Asia' rather than the traditional 'Near East'. The term the Near East has dropped out of usage, submerged in modern political and journalistic parlance by the Middle East, and the Middle East is loosely applied to great areas of Asia. Chapter 1 starts off by defining southwest Asia as a sub-continental region, within which the transformation began and developed. One antique geographical term is essential, however: I continue to refer to the lands on the eastern littoral of the Mediterranean as the Levant, a word that originally referred to the land of the rising sun (when seen from a European or Mediterranean perspective). Within our subject, the Levant (Israel, the Palestinian territories, Jordan, Lebanon, western Syria, and the adjacent part of southeast Turkey) frequently forms a coherent cultural region, and I know no other simple way of labelling that part of southwest Asia.

This book is not simply a statement of the present state of research, although I hope that it tells an up-to-date story. And I do not propose to take the reader through a lengthy discussion of the theoretical debates. The first two parts of the book present a summary of information, the first setting the geographical and biological scenario, the second showing what the archaeology looks like, what it tells us, what the problems have been, and what important discoveries have been made along the way. It is an inconvenient fact that the transformation with which this book is concerned did not happen neatly within the chronological boundaries of the Neolithic period (from 9600 to 6000 BCE). The beginnings of the process are first observed around 23,000 years ago, and there are important developments documented in the archaeology of the final part of the Palaeolithic period, which in southwest Asia has been labelled the Epipalaeolithic. The transformation is better labelled the Epipalaeolithic–Neolithic transformation, therefore; that is such an uncomfortable mouthful that I have generally abbreviated it as the ENT.

The third part of the book presents a new way of understanding this transformation, and a way seeing it as of pivotal importance in human history. One of the features of prehistoric archaeology all around the world has been the expansion of the range of scientific disciplines that have been brought to bear on archaeological questions. Archaeologists have always been borrowers and collaborators, and in the research on the Neolithic, they have devised new hybrid sub-disciplines, such as archaeo-botany and archaeo-zoology (the study of the plant and animal remains from archaeological sites). The accidents of my own career have led me in the particular direction of cultural evolutionary theory which, until recently, had been developing without reference to the archaeology of prehistory. The bundle of disciplines within the field of cultural evolutionary theory are now finding that archaeology can offer hard (that is, cultural material) evidence in support of, or contradicting, theories. Archaeology is also useful for its ability to fix things in an absolute time-scale. As well as helping cultural evolutionary scientists, archaeology can exemplify cultural evolution in process and produce new kinds of explanation of the archaeological record.

The last section of the book sets the period at a pivotal point within the multi-million year long term of human cultural history, setting the transformation within the context of contemporary cultural evolutionary theory. A great deal of speculative and imaginative stuff has been written in recent years, purporting to elaborate on the religious beliefs and practices of Neolithic communities in southwest Asia (just as much has been written over the last hundred years about the religious beliefs that supposedly underpin Neolithic monuments like Stonehenge in southern England). Authors are making reputations – and, presumably, a living – writing, for example, about the Gods and Temples of the extraordinary archaeological site of Göbekli Tepe in southeast Turkey. Having become involved in the arguments, I have interposed a chapter in the final section of the book which seeks to locate the place of the Neolithic within the emerging studies of the cultural evolution of religion.

The book ends with a final chapter that sets out to reverse the perspective with which we have been considering the ENT, the Epipalaeolithic–Neolithic transformation. There is good reason for us to know about the beginnings of the radically new way of life that first emerged around 12,000 years ago; I believe that it represents the initiation of the Anthropocene period. The Anthropocene as a geological period was proposed by climate scientists, biologists, and geologists who had become alarmed by the clear signs of the accelerating impact of human activities on the concentrations of carbon dioxide in the atmosphere, the acceleration in the warming of global temperatures, rising sea levels, and accelerating biodiversity loss (Crutzen & Stoermer 2000). The Anthropocene was the title of a new, 'human-dominated, geological epoch' (Crutzen 2002: 23). I argue in that final chapter that the conditions at the root of the Anthropocene are to be identified in the way that the Neolithic way of life depended on and promoted accelerated population growth, expanding the capacity for cultural innovation, and requiring the continual expansion of the exploitation of Earth's natural resources of all kinds. The Epipalaeolithic–Neolithic transformation is important for us today, because we still depend on the cultural conditions and that were initiated then, and we need to change those characteristics that have been ingrained over more than 10,000 years.

From the start, archaeological research on the ENT, the Epipalaeolithic–Neolithic transformation, has been an international field: there are archaeologists from many countries of the world for whom some aspect of this subject has been the driver of their research. But field research in the twentieth and twenty-first centuries has not moved forward in a simple, straight line. Geo-political events, revolutions, conflicts, and wars have abruptly ended fieldwork projects in first this country and then in that. Those conflicts have also prevented would-be researchers from working where they would most like to direct their investigations. There have been sudden switches of the regional focus of interest, easily explained by the effects of regional politics on the geo-political stage from the 1950s onwards. In consequence, archaeological research in some regions within southwest Asia, the southern Levant in particular, plays a starring role, while other key regions have been underexplored and are scarcely mentioned here.

Since 1970, some of the most interesting and unexpected discoveries have been made in areas that had always been big blanks on the archaeological map. The construction of massive dams on the major rivers of the region, first the Euphrates in Syria, then upstream in Turkey, and then on the Tigris and its tributaries both in north Iraq and southeast Turkey, has meant the drowning and destruction of many ancient archaeological sites. The archaeological and cultural heritage offices of the governments of those countries called for international help to locate sites in the areas behind planned dams and salvage as much archaeological information as possible before enormous areas behind the dams were drowned. Many archaeologists, myself included, have been drawn to conduct their field research by means of salvage excavations; the sites have chosen the archaeologists, rather than the archaeologists have chosen where to work. For different and changing reasons, the course of primary research on the ENT in southwest Asia has been anything but a thoughtfully planned straight line. While we have an uneven database, there is now an extraordinary amount of information; and I hope to show that it makes an extraordinary amount of sense.

Archaeological Terms for Chronological Periods

Prehistory is like history in that it is necessary to get things into chronological order in order to be able to perceive the historical process. Unlike historians, who can name periods after kings, pharaohs, or dynasties, prehistoric archaeologists had to invent labels for their chronological periods, preferably scientific-sounding technical terms. As they have acquired more information that enables them to be more precise, they have had to refine (and complicate) the existing terms. We end up today with a load of chronological jargon that deserves a little explanation. At the end of this section is a (simplified) table of the chronological periods and their approximate dates, as used in this book.

Since the nineteenth century, archaeologists have been spending a great deal of effort identifying and dating the material that they have found. The naming of prehistoric periods followed the example of the early geologists. Thus, the oldest period, the Old Stone Age, the Palaeolithic, coincided with the Pleistocene; and the New Stone Age, the Neolithic, followed in the geological Holocene. The time-span of this book encompasses the last thirteen millennia of the Palaeolithic period (which conventionally ends with the end of the geological Pleistocene around 11,600 years ago, or 9600 BCE), and the Neolithic. Archaeologists working in southwest Asia have adopted the label Epipalaeolithic for those last millennia of the Palaeolithic. They have also devised different labels for the sub-periods that they have been able to define, largely on the basis of the chipped stone tools and the way that they were made at different times and in different regions. For the purposes of this book, I have simplified the chronology of the Epipalaeolithic into two phases, a long early period and a much shorter late Epipalaeolithic, and three phases within the Neolithic. The date-ranges are based on calibrated radiocarbon dates.

The process of Becoming Neolithic requires us to start the account with the developments within the Epipalaeolithic that lead into the Neolithic. So, it is referred to throughout as the Epipalaeolithic–Neolithic transformation, which, for the convenience and comfort of readers, is collapsed into the tag the ENT.

When in the 1950s Kathleen Kenyon began working her way down through the enormously deep stratified remains of Tell es-Sultan, ancient Jericho, she found that there were several metres of deposit that were clearly Neolithic, but which lacked pottery. The Neolithic had been conventionally defined in Britain and Europe in terms of the addition of pottery and other craft-made things to the simpler chipped stone repertoire of earlier times. Kenyon needed a new label for those earliest strata that were in every way Neolithic except for the lack of pottery. She called them Pre-Pottery Neolithic, and, since the strata were clearly divided into quite different earlier and later cultural traditions, she labelled them Pre-Pottery Neolithic-A and Pre-Pottery Neolithic-B. When similar cultural remains were identified at other sites in the southern Levant, they were identified as PPNA or PPNB. Since those labels have come to be used as tags for chronological periods, but also for identifying cultural groups, it can be confusing. So here the terms PPNA and PPNB are avoided, and the chronological sub-periods are called the earlier and the later Pre-Pottery Neolithic. Those periods are followed by a Late (or Pottery) Neolithic, which does have pottery. The table below shows the period names that will be used here, and the approximate dating in years BC (or Before the Common Era).

Period	Date-range
Early Epipalaeolithic	21,000–13,000 BCE
Late Epipalaeolithic	13,000–9600 BCE
Early Pre-Pottery Neolithic	9600–8800 BCE
Later Pre-Pottery Neolithic	8800–6900 BCE
Late Neolithic	6900–5500 BCE

References

Bar-Yosef, O., & Meadow, R. H. (1995). The Origins of Agriculture in the Near East. In T. D. Price & A. B. Gebauer (Eds.), *Last Hunters, First Farmers: New Perspectives on the Prehistoric Transition to Agriculture* (pp. 39–94). Santa Fe, NM: School of American Research Press.

Barker, G. (2006). *The Agricultural Revolution in Prehistory: Why Did Foragers Become Farmers?* Oxford; New York: Oxford University Press.

Childe, V. G. (1936). *Man Makes Himself*. London: C. A. Watts.

Childe, V. G. (1942). *What Happened in History*. Harmondsworth: Penguin Books.

Crutzen, P. J. (2002). Geology of Mankind. *Nature, 415*, 23. doi:10.1038/415023a

Crutzen, P. J., & Stoermer, E. F. (2000). The "Anthropocene". *Global Change Newsletter, 41*, 14–8.

Fuller, D. Q. (2010). An Emerging Paradigm Shift in the Origins of Agriculture. *General Anthropology, 17*(2), 1–12. doi:10.1111/j.1939-3466.2010.00010.x

Peake, H. (1928). *The Origins of Agriculture*. London: E. Benn limited.

Price, T. D., & Bar-Yosef, O. (2011). The Origins of Agriculture: New Data, New Ideas: An Introduction to Supplement 4. *Current Anthropology*, 52(S4), S163–74.

Rindos, D. (1984). *The Origins of Agriculture: An Evolutionary Perspective*. Orlando, FL: Academic Press.

Smith, B. D. (1994). *The Emergence of Agriculture*. New York, NY: Scientific American Library.

Watkins, T. (2010). New Light on Neolithic Revolution in South-West Asia. *Antiquity*, 84(325), 621–34.

1   A Concentration of Opportunity

In this chapter, we need to set the stage and sketch the scenario. Southwest Asia is a region on a sub-continental scale, within which there is a great variety of landscape and climate. One of the purposes of this chapter is to define where within the wider map of Southwest Asia the Epipalaeolithic-Neolithic transformation was generated. The chapter begins, therefore, with a simplified sketch of the physical map of the region complemented by an outline of the present-day climatic variation across the region. For the period with which we are concerned, people's daily lives, their reproduction, and long-term survival or prosperity were very much concerned with the food resources offered by the particular environment within which they lived. Large parts of southwest Asia were endowed with a great diversity of plant and animal species that provided a rich support for mobile foraging societies of the Palaeolithic and Epipalaeolithic periods. Among those plant species were the wild cereals and legumes such as lentils, peas, beans, and chickpeas; these nutritious and storable seeds were valued by the mobile foragers of the Upper Palaeolithic and Epipalaeolithic and began to be cultivated, leading to the domesticated wheats, barley, and legumes of the Neolithic. Southwest Asia also boasted an extraordinary range of animal species for hunters, large, medium, and small mammals; birds; reptiles; and fish, among them were the wild sheep, goat, pig, and cattle that were the first to be domesticated.

This set of relations between population, resources, and environment was a complex and dynamic relationship on two fronts. On the one hand, human populations, like other animal populations, tend to over-reproduce in ways that seek to ensure the reproduction of the population despite loss through infant mortality, disease, and accidents. Scientists can estimate the rate at which a human population would tend to increase if those factors were removed; if we could estimate the actual population density of a region, it might be possible to see how it changes over time. On the other hand, across the period with which we are concerned, there were great climatic changes. Our period begins at the time of the Last Glacial Maximum of the Pleistocene, and it ends several millennia into the Holocene, when climate similar to that of today had become established. Between the Last Glacial Maximum and the establishment of the Holocene climate was not a smooth improvement as temperatures recovered and amounts of annual rainfall (or snowfall) changed. As climate changed, the basic subsistence resources – the plants and

DOI: 10.4324/9781351069281-2

animals from which people could support life – would have changed. This chapter sets out to explore this dynamic relationship between the natural tendency for a human population to grow, the resources that they could obtain from the environment to support life, and the changes in that environment occasioned by climate change.

The Physique, Climate, and Environment

Southwest Asia between 24,000 and 8000 years ago has the same recognisable outline and shape as today's maps but was a series of very different landscapes. We can dispose of the frontiers of the contemporary states, which were drawn and re-drawn in the twentieth century. We can forget about the present-day centres of population, as well as those of historic and ancient historic times. Indeed, we have to imagine a landscape stripped of modern population levels. We should start with a map of the physical shape of the region (Figure 1.1). For readers who have not travelled in the region or thought about its physical geography, the starting point is to recognise that we are looking at a large region at the east end of the Mediterranean that is sub-continental in scale. In total, it is somewhat less than the area of Europe. Both physically and climatically, it is a very varied region.

There are major mountain ranges. Starting from Turkey in the west, there are mountain ranges along the north and south sides of Anatolia, the Asian landmass of Turkey. Between the Black Sea mountains in the north and the Taurus mountains in the south, the centre of Anatolia is a plateau with a base altitude of about 1000 m (3300 ft). The term plateau normally conjures to mind an extensive, flat upland

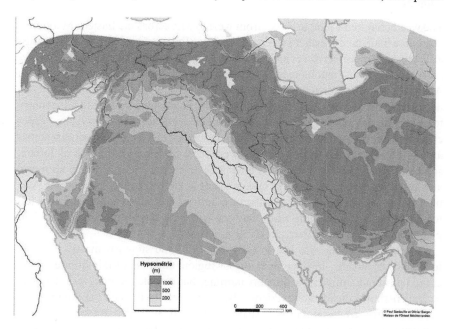

Figure 1.1 Physical map of southwest Asia. (Courtesy of ASPRO, MOM, Université Lyon II)

area, but the Anatolian plateau is often hilly and even mountainous. The Konya plain, where there are some important Neolithic settlements, is at the very centre of Anatolia and is extremely flat; it is the silted up floor of a former, shallow lake that dried out before the end of the Pleistocene period. The eastern half of Anatolia becomes more and more mountainous and rugged. The Taurus chain snakes through the southeast of Anatolia, with many peaks reaching heights in excess of 3000 m, only a little less than the tallest of the Alps in Europe or the Rocky Mountains in North America. The eastern end of the Taurus mountain chain runs north of the modern frontier between Turkey and Syria, swinging southeast to become the Zagros mountains, a chain of more than 1600 km (1000 miles) whose highest peaks top 4400 m (almost 15,000 ft). The frontier between modern Iraq and Iran runs down the Zagros chain. Beyond the Zagros mountains, most of inner Iran is another high, semi-arid plateau. Behind the Mediterranean coast of Syria, Lebanon, and Israel there are lesser mountains and hills running north-south. Those mountains and hill ranges form two main ridges, in Lebanon called the Lebanon and anti-Lebanon mountains. Further south, the divide between the Mediterranean hill country of Israel and the Jordanian highlands is a northward extension of the African Great Rift Valley and the Gulf of Aqaba. Parts of that rift around the Dead Sea are well below sea level. The rift continues north as the Jordan valley. Near its northern end is the Sea of Galilee (also known as Lake Kinneret). Further north again, overshadowed by Mount Hermon and the Golan Heights.

Climate

The weather systems tend to travel from west to east across the Eastern Mediterranean. In winter, cold weather systems from inner Eurasia tend to press southwards, affecting winter temperatures and snowfall, especially in Turkey and Iran. The Aegean coastlands of Anatolia receive a reasonable amount of rainfall, but the central Anatolian plateau tends to be dry, relying on winter rain and snow. The Mediterranean coastland of Syria, Lebanon, and Israel has enough mountain and hill country to ensure fair amounts of annual rainfall from the incoming weather systems, especially in winter. Amounts of annual rainfall reduce rapidly as one moves east across and beyond the Jordan valley or from western to eastern Syria. Much of Jordan, southern Israel, and eastern Syria is technically semi-arid or arid. These semi-arid and arid regions are the northern end of the deserts that occupy much of the Arabian Peninsula. The eastward-moving weather systems have enough moisture left to deliver moderate amounts when they encounter the Zagros mountains and their piedmont hills; but inner Iran quickly becomes arid.

In general, summers are hot and dry, and rain (and snow) is mostly confined to the winter months. In large parts of our region, amounts of rainfall are on average just about sufficient to support farming, but climate records (and historical accounts) show considerable variation from one year to the next, and there have been bad times when rainfall was below average for several years together (as is happening in parts of the region, as global warming is reducing rainfall over recent decades).

Fed by the melting snows of the mountains of eastern Turkey, the great rivers Euphrates and Tigris with their substantial tributaries gather massive flows that pour out of the Taurus mountains into the piedmont and then the plains of north Mesopotamia (another antique term for the lands between the rivers). However, after they have left their narrow valleys in the mountains, these two great rivers have cut down channels with very narrow floodplains as they flow through Syria, for the Euphrates, and northern Iraq, for the Tigris; they do nothing to water the land through which they flow until they come close together in southern Iraq, the heartland of ancient Babylonia. South of modern Baghdad the land between the two rivers is alluvium. All the way to the present head of the Gulf, south of modern Basra, the alluvium is a massive delta region that is squeezed between the Zagros mountains to the east and the limestone shield of the great Syrian desert to the west. In ancient southern Mesopotamia, there was plenty of water and vast amounts of alluvium, but the waters needed to be distributed and spread by irrigation canals and channels. The waters of the Tigris and the Euphrates, draining from limestone mountain ranges, carry dissolved salts that can easily turn areas of over-watered alluvium into saline desert.

Defining the Hilly Flanks of the Fertile Crescent

Field research on the origins of agriculture began in 1948. The American archaeologist Robert Braidwood had become interested in the subject when he was a student at the University of Chicago in the 1930s. He was excited by the ideas of Gordon Childe, the Australian professor of prehistoric archaeology at the University of Edinburgh, who had proposed the idea of a Neolithic or agricultural revolution. Childe, sitting at his desk in Edinburgh, hypothesised that farming in the ancient Near East began in what was called the Fertile Crescent. The idea of the Fertile Crescent was originated by an early twentieth-century ancient historian and archaeologist, James Henry Breasted (1916), who noted that the great civilisations of the ancient world had arisen in a great arc stretching from the Nile in Egypt, up the Mediterranean coastlands, homeland of the Biblical kingdoms, across northern Mesopotamia south of the Taurus, homeland of the Assyrian empire, and southeast through southern Mesopotamia, where first Sumer and Akkad had arisen, before Babylon became the centre of power.

At a time when there was very little palaeo-climate data and no absolute dating methods, Childe had adopted the then current hypothesis that the wetter climate of the Pleistocene for southwest Asia had given way to a much drier regime with the coming of the Holocene period. He therefore reasoned that hunter-gatherers had adapted to these changed climatic times by concentrating around the Fertile Crescent. He had read of the work of an extraordinary American mining engineer and geologist, who in later life had explored in central Asia, and had found and excavated mounded prehistoric sites located in oasis areas where streams from the Kopet-Dag mountains in eastern Turkmenistan run out on the arid steppe (Pumpelly 1908). Childe took the oasis model, substituting the Nile valley and delta, the green Levant and the Euphrates and Tigris valleys for the oases. There, without any archaeological or environmental evidence, he proposed that people found themselves confined by the increasingly arid conditions, spurred to turn to the cultivation of crops of cereals and the herding of the animals that, like the humans, were concentrating in the Fertile Crescent. It was a remarkably resourceful hypothesis.

When Robert Braidwood planned to test and refine Childe's Neolithic revolution theory in the field, he took a geologist colleague, H. E. Wright, with him to look for geomorphological information that might bear on climate and environmental change. Later the team was joined by the Dutch palaeo-botanist Willi van Zeist. Their task was to identify any evidence of late glacial and early Holocene climate change. When Braidwood moved his fieldwork from the piedmont of northeast Iraq to the intermontane valleys of the Zagros in western Iran, his team rapidly found Lake Zeribar and obtained a core from the lake-bed that produced good pollen evidence of the environmental conditions over the period of their interest. Not surprisingly perhaps, they found that the final Pleistocene in the high valleys of the Zagros was a cold treeless steppe. Kinder conditions returned only after the beginning of the Holocene period. It appeared therefore that in that region agriculture had begun in the Neolithic, when warmer, more moist conditions had replaced the arid cold of the final Pleistocene: this result turned the climatic schema of Childe's 'oasis theory' upside-down, but it did not resolve the question that continued to bug Braidwood's research, namely what had triggered the adoption of cultivation of crops and animal herding. And it was clearly unsatisfactory to reconstruct the effects of late Pleistocene to early Holocene climate across the varied physique and environments of Southwest Asia from a single pollen core from a high-altitude lake in a valley of the Zagros mountains.

Braidwood also wanted to identify the environmental zone within which he could reasonably expect to find archaeological sites that would document the transition from hunting and gathering to farming. The plants that were cultivated and the animals herded by farmers in Southwest Asia were wheat and barley, legumes such as peas, beans, lentils, and chickpeas, and sheep and goat, and, to a lesser extent, cattle and pig. When Braidwood asked where the wild ancestors of these domesticated species would have been found in the pristine environment of Southwest Asia at the end of the Pleistocene period, the answer for all of those wild plant species was that they were annuals that were still to be found in the moderately watered hills of the Levant, around the arc of piedmont south of the Taurus mountains, and down the piedmont and intermontane valleys of the Zagros. Wild cattle and wild pig were probably to be found more widely, but wild sheep and goat were definitely animals of the hills and, in the case of the goat, of the mountains. The physical map helped, but the rainfall map (Figure 1.2) made it clear. The palest blue areas in this map are classified as arid, and the next level of pale blue, with less than 200 mm of annual rainfall, is semi-arid. The wild cereals and other grasses, the legumes, the pistachio and almost that could provide nuts, and the landscape that would provide the grazing for wild sheep and goat, lie in that middle blue zone, which is the region that Braidwood identified as 'the hilly flanks of the Fertile Crescent', where there is more than 200 mm of annual rainfall.

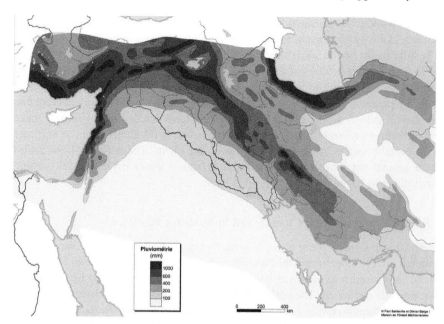

Figure 1.2 Annual rainfall, in mm per annum. The critical area where the greatest number of potential food resources are concentrated has a rainfall above 250 mm per annum, the middle of the range of shaded zones, which coincides with the hill-country of the Levant, the piedmont south of the Taurus mountains in southeast Turkey, north Syria and Iraq, and the Zagros piedmont and intermontane valleys. (Courtesy of ASPRO, MOM, Université Lyon II)

The climate of Southwest Asia is diverse, as one would expect for such a large and varied region. Summers are hot or very hot, and mostly dry; winters are cool, except in the mountains and plateaux, where they can be extremely cold. As well as enjoying moderate rainfall, Cyprus and the east Mediterranean coastlands benefit from the moderating effect of the sea. The defining characteristic of the climate is seasonality. Plants and animals need to be able to cope with long, hot and dry summers, when the rate of water evaporation far exceeds the small amounts of rainfall. In Braidwood's 'hilly flanks' zone the natural environment at the end of the Pleistocene and the beginning of the Holocene period showed that the grasses included wild barley, two species of wild wheat, as well as wild legumes such as lentils, beans, peas, vetches, and chickpeas. These were also the reconstructed habitats of wild sheep, as well as wild goat, gazelle, fallow, and red deer, a now extinct species of wild ass, wild cattle, and pig. For hunter-gatherers the hilly flanks zone offered a remarkable concentration of plant and animal resources. For the first cultivators, the wild cereals and legumes were the focus of their attention; and the sheep, goat, cattle and pig were species that were found throughout most of the hilly flanks and were capable of breeding under human control; they were in effect pre-adapted for domestication.

Braidwood (1960) defined the hilly flanks zone as the region within which people would have begun to cultivate cereals and legumes and to herd sheep and goat. The hilly flanks zone also defined the region within which he had the best chances of finding the archaeological sites that he needed to document the transition from hunting and gathering to the beginnings of farming. Recent information coming from Epipalaeolithic and early Neolithic sites in central Anatolia is indicating that we should extend the arc of the hilly flanks zone to include that region as a place where hunter-gatherers began to harvest and store wild plant foods, to cultivate wild cereals, and to domesticate the local wild sheep. These parts of Southwest Asia, around the arc of the hilly flanks zone, plus a westward branch into central Anatolia, will be the main focus for this book.

Climate, Environmental Change, and Subsistence Resources

We now know much more about climate change between the Last Glacial Maximum of the Pleistocene and the Neolithic period in the early Holocene. The debate about the causal role of climatic and environmental change in the beginnings of farming has continued. The coring of the massive ice-sheet that covers Greenland has allowed environmental scientists to reconstruct and date the pattern of variation in annual temperature over tens of thousands of years before the present. For our purposes, the pattern of climate change between about 23,000 and 8000 years ago is important, and a quartet of British environmental scientists have analysed and synthesised the evidence for the east Mediterranean and the western part of our region of Southwest Asia (Robinson et al. 2006). Palaeo-environmental scientists identify four phases (Table 2.1): our archaeological story begins at the height of the Last Glacial Maximum (23,000–19,000 years BP – before present) within the Pleistocene period. The end of this phase was marked by what is known as Heinrich event, around 16,000 BP, when there were changes in ocean circulation in the north Atlantic that cooled temperatures in the northern hemisphere. In southwest Asia this second phase was a warmer and moister interval, which had also been recognised in the pollen analysis of peat cores in northern Europe. It has been labelled the Bølling-Allerød period and is dated between about 15,000 and 13,000 BP. Pollen analyses from cores in peat bogs and the Greenland ice cores were agreed that, at the end of the Pleistocene, there was a sudden return to cold period with reduced precipitation. It is labelled the Younger Dryas phase and dates approximately 12,700–11,600 BP (10,700–9600 BCE). The Younger Dryas is important to us because there has been much debate about its impact on environments in Southwest Asia immediately before the beginning of the Neolithic period; hence archaeologists usually refer to this period in terms of dates BCE.

Finally, the Holocene period (and our Neolithic) begins around 9600 BCE. Temperatures and annual precipitation recovered quickly, and conditions since then have been relatively (and unusually) stable. Within that relative stability there have been lesser climatic events; two 'rapid climate change' events have been identified within the Neolithic of Southwest Asia, one around 8200–8000 BCE, the other around 6500–6000 BCE. Both involved sudden reductions in annual temperature and reduced annual rainfall. It is best to leave the discussion of the possible impact of these rapid climate change events on the early farming populations to later chapters, when we can set the archaeological picture of settlements and farming practices in the context of a climatic fluctuation.

A recent study carried out by a multi-disciplinary group of scientists (Roberts et al. 2017) has set out (like Robinson 2006) the likely effects on the diverse landscape of Southwest Asia of these climatic phases that have been determined in Arctic ice and north European peat. Roberts and his colleagues were interested in discovering whether there are correlations between environmental change and demographic or cultural change within the ENT, the Epipalaeolithic–Neolithic transformation, our period of interest. In other words, is there any validity in the argument that the beginning of agriculture in the Neolithic was an adaptive response to environmental pressures. They set 'threshold values' for average winter temperature and annual amounts of precipitation: bearing in mind the concentrations of key plant-food resources for the Epipalaeolithic hunter-gatherers, which included the potential domesticates used by the first cultivators, their rainfall threshold was at least 240 mm per year, and their average winter temperature threshold was no colder than two degrees lower than current winter temperatures; their estimates for both values are tabulated in Table 2.1.

They believe that the average annual rainfall for the Last Glacial Maximum was only 60% of twentieth-century annual rainfall, and winter temperatures were seven degrees Celsius lower than the present. The Last Glacial Maximum was certainly the greatest departure from recent experience for those living in the region. The two-millennia-long Bølling-Allerød warm phase for Southwest Asia brought increased levels of precipitation (30% above twentieth century levels) and winter temperatures that were only about one degree Celsius colder than present. The Younger Dryas phase represented a sudden return to glacial conditions in the Arctic; in their model for Southwest Asia Roberts et al. estimate that mean winter temperatures were four degrees colder than present, and mean annual precipitation was reduced to only 75% of recent times. Finally, the rapid transition from the Younger Dryas into the early millennia of the Holocene brought winter temperatures that were a little warmer than recent experience, and considerably greater annual precipitation (again, about 30% higher).

The next stage in their work was to model those generalised figures for Southwest Asia into maps that would show which parts of the region were affected by environmental conditions that were above or below those threshold conditions. Robert et al. provide a reference map of current or recent conditions against which we can compare the maps for the earlier periods. The Last Glacial Maximum squeezed environmental conditions severely, of course, but nevertheless much of Braidwood's hilly flanks of the Fertile Crescent survived. On the east coast of the Mediterranean, the viable strip was narrower but still included the Jordan valley and perhaps a strip of the high ground on its eastern rim. Around the northern arc of the hilly flanks zone, through northeast Syria, southeast Turkey and across north Iraq to the Zagros piedmont, the territory that was above the threshold values was narrow and may have been fragmented.

The map for the Bølling-Allerød phase is quite similar to the present-day map. More of the mountainous east of Turkey was worse than the threshold conditions, as was a strip of the mountains behind Turkey's Black Sea coast. Like the area around Lake Zeribar, high in a Zagros mountain valley, these mountainous regions were slow to recover from the glacial period. The Younger Dryas phase map is important for us, because the eminent archaeo-botanist Gordon Hillman and some archaeologists have argued that the adversity of that sudden cold phase stimulated hunter-gatherers in the Levant to begin to cultivate wild cereals in response to the reduction in the availability of their traditional wild resources (e.g. Bar-Yosef & Belfer-Cohen 1992; Moore & Hillman 1992). This hypothesis became the standard view (though we shall see later that it was being questioned by some). The only sources of information of primary importance to these arguments for the harsh impact of the Younger Dryas and its causal role in requiring Epipalaeolithic hunter-gatherers to begin cultivating cereals have been two pollen cores from shallow lakes in the Levant, and the archaeo-botanical data from the late Epipalaeolithic settlement at Abu Hureyra in north Syria, which will figure in the next chapter. There are also lake-bed cores from lakes in Anatolia, including Lake Van in the far east of Turkey; but the lakes are at relatively high altitude that were slow to recover from the Last Glacial Maximum, and they do not represent conditions in the hilly flanks zone. The pollen cores from Ghab in Syria and the former Lake Huleh in the far north of Israel have always been problematic, and debate continues over the significance of shifts in the pollen diagrams and the radiocarbon dating. In the publication of the most recent work, based on more cores, the researchers review the problems arising from earlier cores, radiocarbon dating and interpretation; in their own work, they are concerned only with the Holocene environment, effectively abandoning the final Pleistocene and the Younger Dryas phase (van Zeist et al. 2009).

Robinson et al. (2006) who have tried to put together all the relevant evidence from different sources concluded that the Younger Dryas was colder and very arid, but they also remark their surprise that there is not a clear signal of the Younger Dryas to be found in Eastern Mediterranean marine record. Environmental reconstruction from pollen cores is based on the counting of tree pollen grains and the changing ratios among tree species and between trees and plants associated with cold or dry conditions. Hillman's interpretation of the impact of the Younger Dryas modelled a narrowing of the Mediterranean woodland zone in the Levant (Moore & Hillman 1992). He extrapolated a die-back of the open woodland across the northern arc of the hilly flanks zone and worse effects further east. While some have relied on these pollen diagrams to show that the Levant suffered a fairly severe reversal during the Younger Dryas, others have found locally contradictory evidence, and others again have concluded that the evidence is equivocal. The site of Körtik Tepe is one of several in the upper Tigris valley in southeast Turkey that is particularly significant. The settlement was established in the middle of the Younger Dryas phase, and it continued through the early centuries of the early Pre-Pottery Neolithic. Based on the botanical assemblage recovered from the excavations, the archaeo-botanists conclude that the trees of the open woodland environment were reduced in the colder winters and more arid conditions of the Younger Dryas; but the array of gathered food plants suggest that the opening up of the woodland allowed a richer ground cover. As woodland tree species recovered in the early Holocene, affecting the plentiful supply of plant resources, the people of Körtik Tepe shifted the balance of their foraging to take a greater interest in riverine species. The advantage of the Körtik Tepe botanical assemblage is that it offers good charcoal evidence of the local availability of tree species, but it also includes direct evidence of many of the locally available plant-food resources that are not represented in pollen cores taken from lake beds. The most recent excavation evidence is that Körtiktepe is not unique in being a settlement that was established during the Younger Dryas phase.

Roberts and his colleagues model the effect of their threshold values for the Younger Dryas, showing that the Levantine coastlands remained above those thresholds, together with a rather slimmer portion of the northern arc of the hilly flanks zone and the Zagros piedmont. Anatolia, except for its Mediterranean and Black Sea littorals and the Aegean west, however, was too cold. The map of the early Holocene shows that the regions that were above the threshold values were somewhat more extensive than those of the twentieth century; it was a climatic optimum that was very similar in temperatures to those of recent times, but with somewhat higher rainfall.

Population

Demography is going to become important later in this book, so it is useful to get basic information together at this stage. Human populations depend for food and for their tools and equipment on resources obtained from the environment. In the remote past, in the Palaeolithic and the Neolithic periods, people obtained the raw materials from which they made the tools, weapons, and equipment that supported everyday life, whether that was stone (flint, chert, obsidian for making chipped stone tools, hard stone for grinding stones, mortars, and pestles), wood, bone or antler. They also used all sorts of vegetable materials for making baskets, string, twine, nets, and matting, but we only very rarely find any trace of such materials. People also needed to obtain various resources, whether from plants or from animals, that they could turn into food (the subsistence economy). There was always a dynamic equation between population and resources, in that both sides of the equation were complex and either could change. Climatic and environmental change might reduce the regional availability of food resources; equally a local surge in population might put pressure on the available resources. It has been customary, since the time of Gordon Childe to think in terms of climatic and environmental changes putting pressure on human populations to adapt by innovation (farming, for example). The picture is a good deal more complex, and an important element of that complexity is demographic. An important paper by Roberts et al. (2017) put the profiles of climatic and environmental change together with the demographic profile of the region in order to assess whether demographic changes coincided with climatic events that might be identified as causal. The title of their article, referring to human responses and non-responses to climatic variations, advertises that they find the relationship between climatic and environmental change and demography was not simply deterministic.

Assessing population levels in prehistory is notoriously difficult. In recent years, there has been some success using the human remains from European Neolithic cemeteries in showing that there was a Neolithic 'demographic transition', the equivalent of the demographic transition associated with the industrial revolution (Bocquet-Appel 2002; Bocquet-Appel & Bar-Yosef 2008). That demographic transition occurred where a population with high birth and mortality rates changed to low birth rate and low mortality. At first, the high mortality rate declined, but birth rates remained high, resulting in an explosive growth in population; later, birth rates declined and the population stabilised as a low birth rate, low mortality rate society. Bocquet-Appel showed that a similar demographic transition could be documented in Neolithic Europe, where the expansion of Neolithic colonists who buried their dead in cemeteries allowed physical anthropologists to profile their birth and mortality rates. The signal of a major demographic shift has been detected in regions, such as Europe, where indigenous hunter-gatherer populations are effectively replaced by the arrival of Neolithic colonists in the archaeological record. By contrast with pre-Neolithic hunter-gatherer groups, there is a relatively abrupt increase in the proportion of immature skeletons in the cemeteries of newly arrived Neolithic colonists, who at first experienced lower mortality and a high birth rate. In the Neolithic of Southwest Asia, there are plenty of burials, but we do not find such customs of systematic burial of the dead in cemeteries. In any case, the situation in southwest Asia is quite different: as we will see in the following chapters, the process of transition from mobile forager bands to sedentary villagers living by farming was long and gradual, so there is no rapid Neolithic demographic transition.

Nigel Goring-Morris and Anna Belfer-Cohen (2010) used archaeological sites and settlements as a proxy for human population numbers. Their study counted the number of sites, period by period, for different regions within Southwest Asia. Their best data comes from the southern Levant, which has seen the greatest amount of fieldwork over more than a century. Despite the probability that a single mobile hunter-gatherer group might be represented by several seasonal encampments, and ignoring the fact that Neolithic settlements represented populations of many hundreds or some thousands, their study showed a steady increase in the number of sites per thousand years from before the Last Glacial Maximum and the beginning of the Epipalaeolithic period. The curve on their graph for sites/population in the southern Levant shows no interruptions that might be attributed to the impact of adverse climatic periods, the Younger Dryas in particular.

Roberts et al. (2017) used a different approach, assembling radiocarbon dates, and plotting the clusters against the timescale: the more people there were in a region, the more sites there would be for archaeologists to find, and the more radiocarbon dates archaeologists would generate. This is the simple assumption behind what is called the summed probability distribution of radiocarbon dates; the denser the radiocarbon dates cluster, the greater the number of sites, and the denser the human population. The assumption may be simple, but the methodology involves sophisticated measures to ensure reliability that do not need to concern us here. Their data covered the period from 17,000 BP, so it does not include the Last Glacial Maximum and the beginning of the Epipalaeolithic period. The study sub-divided the data so that they could produce graphs for three sub-regions, the southern Levant, the north Levant together with the northern arc of the hilly flanks, and south-central Anatolia, where a number of sites have been under recent investigation.

The curves produced from the radiocarbon dates were certainly not smooth, which is what would be expected if a stable population were able to reproduce through time without external interventions of any kind. The real interest is in the overlay of the regional graphs of summed probability distributions, representing periods of population growth or stability or serious external pressure, on the sequence of climatic periods. The graphs for each of the three sub-regions are significantly different. For the southern Levant, the graph shows significant population growth at the time of the Bølling-Allerød interstadial (approximately 14,500–11,500 BCE). This coincides with the already recognised expansion in the number (and size and permanence) of late Epipalaeolithic Natufian sites. This population expansion is followed by further growth in the Younger Dryas period, and accelerating growth through the final part of the Younger Dryas and in the early centuries of the Holocene/early Pre-Pottery Neolithic. This contradicts the hypothesis that the cold and arid Younger Dryas reduced the availability of plant-food resources, pressuring the Epipalaeolithic population, pushing some communities to abandon their sedentary way of life and scatter in a more mobile way of foraging.

We shall look in the next chapter at the claim of cultivation as an adaptive response to the severity of the Younger Dryas phase; the idea that late Epipalaeolithic Natufian communities of the southern Levant were forced to abandon their sedentary way of life has been subject to erosion for some time, as a number of late Natufian settlements have been recently excavated (producing the radiocarbon dates that have boosted the population curve through the Younger Dryas phase). Whereas we used to be told that the Younger Dryas provoked a reduction and dispersal of the late Natufian population, recent archaeological evidence is now confirmed by this study in showing that the population continued and indeed continued to grow during the Younger Dryas, reaching a climax in the late tenth millennium BCE.

The population curve for the north Levant and the northern arc of the hilly flanks zone starts with a long flat line until it reaches the Younger Dryas phase. That simply reflects the near complete absence of known earlier Epipalaeolithic sites across that region. By contrast with the southern Levant, where there is a history of nearly a century of dedicated Palaeolithic research, there has been recent salvage archaeology work in the valleys behind new dams. The earliest sites discovered are Neolithic, with some sites being found to have been established in the last millennium (the Younger Dryas phase) of the Epipalaeolithic period. The curve begins to take off in that Younger Dryas phase, and it continues a consistent upward trend into the Holocene and the early Pre-Pottery Neolithic. Indeed, in the early Pre-Pottery Neolithic the population curve is significantly above expected levels.

The authors estimate that over that period of roughly 2000 years regional population grew approximately twenty-fold. After about 8500 BCE, the population curve declines, corresponding to the reduced numbers of settlements dating to the later Pre-Pottery Neolithic. What lay behind this dip in the curve of summed probability distributions? Did the population suffer a serious decline, or was there a dispersal? These are not questions that can be easily resolved, but they will be considered in a later chapter. To anticipate that discussion, we can note here that another study (Borrell et al. 2015) of material culture from sites of this time between 8500 and 8000 BCE has noted the sudden decline in sites and a cultural hiatus: the sites dating after this period show a marked (at least to expert eyes) cultural discontinuity with the earlier sites.

The third sub-region is south-central Anatolia, which has produced a very different curve from the other two sub-regions. There is a small blip of population in the late Bølling-Allerød phase, corresponding to a late Epipalaeolithic presence (in a single trench at a single site) on the plateau. The Younger Dryas phase is blank, which accords with the reconstruction of the environment at that time as very cold, arid and unwelcoming. On present evidence, the summed probability distribution of radiocarbon dates only begins to pick up about half a millennium into the Holocene period. Between 8300 and 8100 BCE there is a rapid increase in population. Roberts et al. note that this sudden and rapid growth in population coincides with the decrease in population in the north Levant/northern arc of the hilly flanks sub-region, and they speculate whether there was a population migration. Having worked at the site on the edge of the Konya plain that produced the Epipalaeolithic 'blip' followed by the earliest dates for settlement in the Pre-Pottery Neolithic period, I do not see cultural traditions in the Pre-Pottery Neolithic settlements that could be seen as evidence of population movement. I believe that it is premature to think that we have the full settlement history of the Pre-Pottery Neolithic for this sub-region; the earliest sites known to date are still under investigation, and new sites are being discovered.

In summary, then, it seems that we now have a picture of the two dynamic elements in the equation that links climatic and environmental change, with the expansion of population in Southwest Asia. Between these two elements is the central issue of subsistence resources and the strategies that were employed to obtain and process them; we will turn to those matters in the next two chapters. But, at the end of this chapter, it deserves emphasising that population levels began to rise rapidly after the Last Glacial Maximum, with the onset of the warmer, moister Bølling-Allerød phase. Contrary to what has been generally believed, however, the population continued to rise significantly through the Younger Dryas phase, wherever we have evidence. The picture we get from central Anatolia is somewhat different: there, we have begun to perceive the presence of complex hunter-gatherers at the end of the Bølling-Allerød phase, but there was then a blank until the population appeared and began to rise very rapidly several centuries into the Holocene period. The surprise is that the Younger Dryas phase, far from restricting population around most of the hilly flanks zone saw significant growth of human population, which continued into the early Pre-Pottery Neolithic. We will need to return to the subject of population later in the book, when we are looking at the large, sedentary communities who were increasingly dependent on farming in the later Pre-Pottery Neolithic and the late Neolithic; but now we are ready to look at the subsistence strategies across our period in the next two chapters.

References

Bar-Yosef, O., & Belfer-Cohen, A. (1992). From Foraging to Farming in the Mediterranean Levant. In A. B. Gebauer & D. T. Price (Eds.), *Transitions to Agriculture in Prehistory* (pp. 21–48). Madison, WI: Prehistory Press.

Bocquet-Appel, J.-P. (2002). Paleoanthropological Traces of a Neolithic Demographic Transition. *Current Anthropology*, 43(4), 637–50.

Bocquet-Appel, J.-P., & Bar-Yosef, O. (Eds.). (2008). *The Neolithic Demographic Transition and Its Consequences*. Dordrecht: Springer.

Borrell, F., Junno, A., & Barceló, J. A. (2015). Synchronous Environmental and Cultural Change in the Emergence of Agricultural Economies 10,000 Years Ago in the Levant. *PLoS ONE*, 10(8), e0134810. doi:10.1371/journal.pone.0134810

Braidwood, R. J. (1960). The Agricultural Revolution. *Scientific American*, 203(3), 130–52.

Goring-Morris, A. N., & Belfer-Cohen, A. (2010). "Great Expectations" or the Inevitable Collapse of the Early Neolithic in the Near East. In M. S. Bandy & J. R. Fox (Eds.), *Becoming Villagers: Comparing Early Village Societies* (pp. 62–77). Tucson: University of Arizona Press.

Moore, A. M. T., & Hillman, G. C. (1992). The Pleistocene to Holocene Transition and Human Economy in Southwest Asia: The Impact of the Younger Dryas. *American Antiquity*, 57, 482–94.

Pumpelly, R. (1908). *Explorations in Turkestan (2 vols.)*. Washington: Museum of Natural History.

Roberts, N., Woodbridge, J., Bevan, A., Palmisano, A., Shennan, S., & Asouti, E. (2017). Human Responses and Non-Responses to Climatic Variations during the Last Glacial-Interglacial Transition in the Eastern Mediterranean. *Quaternary Science Reviews*, 184, 47–67. doi:10.1016/j.quascirev.2017.09.011

Robinson, S., Black, S., Sellwood, B., & Valdes, P. (2006). A Review of Palaeoclimates and Palaeoenvironments in the Levant and Eastern Mediterranean from 25,000 to 5000 Years BP: Setting the Environmental Background for the Evolution of Human Civilisation. *Quaternary Science Reviews*, 25(13–14), 1517–41.

van Zeist, W., Baruch, U., & Bottema, S. (2009). Holocene Palaeoecology of the Hula Area, Northeastern Israel. In E. Kaptijn & L. P. Petit (Eds.), *A Timeless Vale. Archaeological and Related Essays on the Jordan Valley in Honour of Gerrit Van Der Kooij on the Occasion of His Sixty-Fifth Birthday* (pp. 29–64). Leiden: Leiden University Press.

2 Changing Subsistence Strategies
Foraging to Farming

It is logical to move from the geography of southwest Asia and climate and environmental change in the previous chapter to the changes in the ways that people obtained their subsistence resources within those environments in this and the next chapter. But I am not suggesting by that sequence of chapters that the changes in subsistence strategies across the ENT were responses to climate or environmental change. Nor do I want to imply that the cultural changes that we shall be discussing in the following chapters were consequential on, or driven by, economic changes. How all this information fits together and how it should be understood must wait until the last chapters; reading a book is necessarily a linear experience, but the sequence of chapters is certainly not predicated on an unfolding logic that starts with the causes and goes on to the consequences.

Before we go any further, I want to make it clear that the approach taken here in this chapter about plant foods and the next about hunted or herded animals is rather different from what is commonly found. For a long time, at least since Lewis Morgan's (1877) *Ancient Society*, hunter-gatherers were seen as 'savages', while farming societies were classed as barbarian. Gordon Childe, arguing for the significance of his 'Neolithic revolution', claimed that the beginnings of farming represented the release from savagery. Once plants and animals had been domesticated, man – it was always man – had proclaimed his mastery over nature. For most people, looking back into a deep and dark prehistory it was enough to be told where and when the domestication of plants and animals took place; at that range it seemed like an event that could be given a date and an identified placer on the map. But domestication is a process, not an event; and farming is not some discovery that, once someone had thought of it, could simply replace hunting and gathering. For those prehistoric peoples, the discovery of farming was not a long-term objective towards which they clumsily stumbled over many centuries. A multi-disciplinary group of co-authors have recently argued the case for regarding domestication as a complex dynamic process much better than I could (Weide et al. 2022).

In the next chapter we will be looking at the meat component in the diets of the people who lived around the hilly flanks of the Fertile Crescent across the critical time-period from the Last Glacial Maximum, around 22,000 BP, to the end of the

DOI: 10.4324/9781351069281-3

Neolithic, around 8000 BP. In this chapter we are concerned with the contribution of plant foods, and, most importantly, how those plant-food resources were obtained. At the start of our time-spectrum, all plant-food resources were gathered; and at the end of period, most plant-food resources were farmed. As we saw in the last chapter, the arc of the hilly flanks zone, plus an extension into central southern Anatolia, has been defined as the common habitat of the potential domesticates, those (wild and naturally occurring) species that were pre-adapted to being taken under human control and managed for human use.

Tracing the physical remains of plant-food resources in the past has been beset with difficulties. Why should we expect to find traces of plant foods on archaeological sites? How can excavators hope to recover minute fragments of plant foods, if they have somehow survived? For the Epipalaeolithic period there are flint sickle blades with gloss produced by abrasion against the silica-rich stems of grasses, cereals, sedges or rushes, grinding slabs and mortars and pestles; but there is a serious deficiency in the hard evidence of plants that were harvested and processed. One reason for the lacuna in evidence is that plant materials, including seeds, decay in the soil matrix of archaeological sites, unless they have been carbonised by fire. People learned that it was a good idea to dry harvested grains or seeds before storing them in bulk, and the accidentally carbonised seeds and associated parts are the main, but obviously biased, source of our information. Where carbonised plant remains were incorporated into accumulating archaeological deposits, as in a Neolithic settlement consisting of mud-brick or pisé buildings that required frequent maintenance and periodic replacement, they may survive indefinitely. The small, fragile charcoal fragments need to be buried below the superficial soil where roots, insects, and soil chemistry are active. Thus, carbonised plant remains are unlikely to survive on hunter-gatherer camp sites, which were places of successive, short-term occupations that generally did not accumulate much archaeological deposit and were subject to erosion by wind and rain.

The second reason for the poor evidence for plant use is that techniques for the systematic retrieval of carbonised plant remains were devised and began to be implemented only in the 1970s; and systematic sampling with flotation machines has been rather slowly adopted among field research teams. Wet sieving and flotation began to be developed as an extension of the dry sieving of archaeological deposits. The flotation machine consists of a tank full of (turbulent) running water, and a fine mesh onto which amounts of archaeological deposit are tipped; the fine sediment passes through the mesh, and the turbulence of the water causes the light, carbonised plant materials to float to the surface, where it overflows a weir to be caught in a fine mesh. It is easy to collect the carbonised grains of cereals and fragments of wood charcoal, but archaeo-botanists have become interested in capturing the smallest fragments of chaff, the tiny rachises that connected the grain to the head of the plant, and the often tiny seeds of the other species that were growing among the cereals. And those small seeds are great importance if the site was used in the Epipalaeolithic and early Pre-Pottery Neolithic, when we have learned that many small-seeded grasses and other plants were harvested. The data with which the archaeo-botanists have to work, therefore, is very uneven: we obviously only know of those species whose seeds were accidentally carbonised; on some sites the excavation trenches may have been in the wrong locations and have missed the places where the ash from the hearths and the accidentally carbonised seeds were thrown.

The Critical Importance of the Epipalaeolithic

Although very few of us today have any direct experience of farming, it seems entirely natural to us, whether in terms of the 'agriculture industry' that is the ultimate source of our food today, or in the romantic views of rural life that we get from nineteenth century novels, poetry and painting, when the old ways were dying. We can therefore think that it is strange that our so-called wise species, Homo sapiens, which had been around for about 300,000 years, only began to turn to farming around 10,000 years ago. Most of us are even more unfamiliar with the way of life of hunter-gatherers. From our point of view farming has so many obvious advantages; on the other hand, living in small bands of mobile foragers, moving repeatedly throughout the year from camp-site to camp-site, reliant on what can be found out there, seems a strange and precarious way of life. As we have seen, Gordon Childe set the tone, boldly proposing that climate change – aridification of much of southwest Asia – at the end of the Pleistocene demanded the change and presented the opportunity (Childe 1936).

It is not a story that no longer inspires confidence. Robert Braidwood, whom we met in the previous chapter, rejected the idea that climate change played a significant role in pushing people to begin farming. He hypothesised that the transition from mobile hunting and gathering to farming needed two intermediate stages: intensified plant-food collection, leading to incipient cultivation, out of which dependence on domesticated crops finally emerged. Unfortunately, although he led programmes of survey and excavation in northeast Iraq, western Iran, and finally in southeast Turkey, he was unable to test of document this hypothetical four-stage transition process in his own results. Although he continued to work within the Neolithic period, he had pointed to the importance of the (Epipalaeolithic) period that led the way to the Neolithic.

In the late 1960s two young American archaeologists, who both had worked with Braidwood, and who were leaders in a movement to upgrade anthropological and archaeological theory by setting prehistoric change in the context of ecological systems and evolutionary theory. In a book that announced 'new perspectives in archaeology', Lewis Binford contributed a chapter that proved to be very influential (Binford 1968). His essay was concerned to propose a new way of approaching the emergence of agriculture, in which populations were seen as being either in equilibrium or in disequilibrium with their environment and its subsistence resources. If the human population component in an ecosystem was in disequilibrium, then some kind of adaptation would follow. Binford proposed that, at the end of the Pleistocene period, world sea levels rose as polar ice and glaciers melted, pushing human groups living close to the sea to move back into land where there were already other groups. Hence, the end of the Pleistocene led to the post-Pleistocene adaptation of agriculture to relieve the demographic stress (Binford 1968: 328).

Kent Flannery and Frank Hole, who had been working with Braidwood in western Iran, had undertaken a programme of their own field research in southwest Iran. Flannery effectively answered Binford's challenge when he proposed his broad-spectrum revolution theory at a conference in London on the domestication and exploitation of plants and animals (Flannery 1969). His hypothesis of a broad-spectrum revolution in the Epipalaeolithic of Southwest Asia was based on the fieldwork that he and Frank Hole had carried out in southwest Iran, divided between Palaeolithic sites in the Khorramabad valley of the Zagros mountains and Neolithic sites on the adjacent Deh Luran plain of southwest Iran (Hole et al. 1969; Hole & Flannery 1968). They found Middle Palaeolithic and Epipalaeolithic sites in the high mountain valley, but nothing Neolithic; down on the alluvial Deh Luran they found a cultural sequence that began with an early Neolithic (it had no pottery in its earliest phase) and continued on through the Neolithic and beyond with more and more mounded settlements.

Flannery's hypothesis of a broad-spectrum revolution said that was not the adoption of farming, but the critical step was taken when Epipalaeolithic hunter-gatherers extended the spectrum of animals and plants on which they relied for food. In terms of animals, broadening the spectrum meant investing more time and skill in capturing small mammals, birds, amphibians, reptiles (tortoises and snakes). In terms of plants, they began to invest more in harvesting storable seeds (grasses and wild cereals) and legumes (lentils, peas, vetches and the like). The broad-spectrum strategy went hand in hand with decreased mobility: Epipalaeolithic groups stayed longer in one place, and, as a consequence, their females tended to have a lesser spacing between births, by contrast with the typical four-year birth spacing among highly mobile hunter-gatherers. Thus, the consequence of the broad-spectrum subsistence strategy was an upturn in the rate of population growth, which, by the end of the Epipalaeolithic period, could no longer support the greater numbers. Therefore, some groups had to leave the valley environment and seek new territory. The seasonal marshy patches on the Deh Luran plain gave them the space that they needed, but it was necessary to bring some of the resources on which they had depended, namely cereal grain to plant in the alluvial soil and sheep and goat whose grazing could be managed and whose reproduction could be controlled. In short, they began to cultivate crops and herd animals in the new environment, and the beginnings of farming were the inevitable consequence that followed from the adoption of the broad-spectrum settlement and subsistence strategy. The scenario for Flannery's broad-spectrum revolution are closely related to Binford's hypothesis, in that both involve people making adaptations in response to increasing population pressure on available wild resources, or, in Binford's case, available land in the face of world sea level rise.

Before we follow the continuing impact and development of Flannery's broad-spectrum revolution theory in the second part of this chapter, we should add into the picture contemporary developments in thinking about hunter-gatherer societies and their settlement and subsistence strategies. A group of anthropologists and ar-chaeologists interested in hunter-gatherer societies published the papers from a conference Man the Hunter (Lee & DeVore 1968). The main point that archaeolo-gists took away from the publication came from two complementary papers. In one Richard Lee showed that the ethnographic record showed that hunter-gatherers spend only a limited time in subsistence tasks, and have a good deal of leisure time (Lee 1968). This evidence knocked out the idea that prehistoric hunter-gatherers were locked in a perpetual struggle for survival, so that escape to the food-producing strategy of farming represented salvation (Lee 1968). The other chapter was by Marshall Sahlins, in which he proposed that prehistoric hunter-gatherers were, in his words, 'the original affluent society', in that they had all that they needed to sustain a rich life (Sahlins 1968). Sahlins followed up with his short book on *Stone Age Economics* (Sahlins 1972), which was widely influential among archaeologists.

A few years later, three papers were published very close in time to each other; together, they defined an important distinction between different kinds of hunter-gatherers with different settlement and subsistence strategies. These distinctions between different ways of living as hunter-gatherers are important to us, both in terms of their subsistence strategies, and the settlement strategies that went with them. Binford expanded on the difference between the Inuit Nunamiut, whom he had been studying, and whom he described them as being 'logistically organised' collectors, and the San Bushmen of southern Africa, whom he described as forag-ers (Binford 1980). From the ethnographic information, he set out to show how the different ways in which these two settlement and subsistence strategies map onto resources in the landscape open the way to understanding prehistoric hunter-gatherer sites. Binford's foragers move frequently from one seasonal patch to an-other, going out each day from their encampment to hunt or to gather, bringing back and consuming whatever they have obtained. Logistically organised collec-tors move their base-camp infrequently; each base-camp location is best situated to allow access to a range of diverse resources in adjacent ecological zones. In the light of Hole and Flannery's Epipalaeolithic base-camp sites in the Khorramabad valley and Flannery's broad-spectrum revolution hypothesis (and what we shall encounter in the next chapter concerning Epipalaeolithic sites and their archaeol-ogy in the Levant), the implication is that the first sedentary communities and the beginnings of cultivation practices arose among societies that Binford would clas-sify as 'logistically organised' 'collectors'.

Taking a different perspective from Binford, the anthropologist James Wood-burn differentiated hunter-gatherers based on whether they engaged in 'delayed-return' or 'immediate-return' strategies (Woodburn 1982). Binford's foragers who brought back to base what they had hunted or gathered that day were Woodburn's 'immediate-return' hunter-gatherers; and the logistically organised collectors were often the hunter-gatherers involved in delayed return. Delayed return might be characterised by investment in technical facilities such as, for example, boats or nets used in fishing, or by the storing of food materials for future use, such as the smoking and drying of fish, or the harvesting of grasses and wild cereals, or the tending and maintenance of trees or other plants, or the management in some form of wild animal herds.

The third paper was by the French anthropologist Alain Testart, who noted that hunter-gatherers who engage in harvesting and storing food supplies for future use are much more like farmers than Woodburn's immediate-return hunter-gatherers (Testart 1982). While small-scale, mobile foraging band societies rely on flexibility and multiple alternative strategies for immediate return, the economies of the sedentary hunter-gatherers such as the Pacific Northwest Coast and California peoples were based on large-scale seasonal food storage, which was partnered by more complex social organisation. The implication for us is that Epipalaeolithic hunter-gatherers in Southwest Asia who had taken up delayed-return strategies involving storage and a more sedentary way of life needed only a few relatively small steps, such as beginning to cultivate crops, to precipitate them towards a farming economy.

As we shall see in the Chapters 4 and 5, parts of Southwest Asia, particularly the hilly flanks zone, were a rich environment for prehistoric hunter-gatherers. This was true not only of the variety of herds of grazing animals, but also of richly nutritious, storable plant foods. A large number of these species, namely the cereals (einkorn and emmer, which are species of wild wheat, barley, and, to a lesser extent, rye and oats), and legumes (peas, horse-beans, lentils, chickpeas and some vetches), while they were not equally universal, were mostly widely available. There were also nuts and fruits to be gathered, although they may rarely leave traces in the archaeological soil. The natural habitats of all these species of cereals and pulses have been reconstructed and mapped (Harlan & Zohary 1966; Smith 1994); they tend to overlap each other (and coincide with the habitats of many of the hunted animal species, including the potential domesticates). Together they practically define the hilly flanks zone, with something of an extension into central Anatolia. Most of these species grew wild, often in dense stands, in the open oak and pistachio woodland.

The general idea grew up that the late Epipalaeolithic use of plant foods at sites in the southern Levant constituted a sort of prelude to the domestication of cereals that emerges in the Pre-Pottery Neolithic. For a long time, the primary interest of researchers was to find the earliest date at which domestication was identifiable and to define the place where domestication occurred, the cradle of agriculture, or the beginnings of farming. Advances in archaeo-botanical research and the analysis of materials from the Epipalaeolithic have broadened that focus. And we have learned that it is false to imagine that the domestication of these species occurred at one place, or at a single moment. Domestication is a process, and defining the beginning, or the completion, of that process is not possible. When we have reviewed some of the evidence that has been found over recent decades, we will be able to trace the story of changing plant-food use over the ENT.

The discovery and exploration of an extraordinary site dating to the Upper Palaeolithic-Epipalaeolithic boundary, around 23,000 years ago during the Last Glacial Maximum period, has disposed of that assumption. The site of Ohalo II was found in 1989 when the summer water level of the Sea of Galilee (Lake Kinneret) dropped to an unprecedented low (Nadel 2017; Nadel & Werker 1999). The whole of the rift valley has been subject to active tectonics, and water levels have risen and fallen repeatedly. Ohalo II must have been occupied at a time when the Sea of Galilee was somewhat smaller than today. Following the Last Glacial Maximum, the site has been submerged in shallow water and fine lake sediments, and the anaerobic conditions have preserved plant remains in uniquely great quantities, whether carbonised or not. We will hear more about the remarkable site of Ohalo II in the next chapter when we are considering the site's faunal remains, but here it is useful to know that, in addition to the extraordinary array of plant remains, the group that occupied the site exploited a wide range of resources that were available across the different ecological zones around the site. Together, the seeds, fruit stones, bird, fish and animal bones indicate that people were there at all seasons of the year, and they must have stayed at the site for considerable periods in order to consume all their harvested seeds and grains. Perhaps the group sometimes spent the whole year there. The plant remains recovered at Ohalo II amount to more than 90,000 seeds representing some 142 taxa; about 19,000 were grass grains, mostly small-seeded species, but including more than 2500 grains of wild barley and 102 wild emmer wheat grains (Kislev et al. 1992; Weiss et al. 2004, 2008). There were also some oats, goat-grass, almond, hawthorn, pistachio, olive and grapes (Kislev et al. 1992).

There is more information on the harvesting and processing of the seeds that have been found at Ohalo II. A study of the small flint sickle blades has shown the gloss that is typical of cutting the silica-rich stems of cereals and grasses. Experimental harvesting of different species at different stages of ripeness has allowed the team to conclude that the group were harvesting near-ripe semi-green wild cereals, before their grains were ripe (Groman-Yaroslavski et al. 2016). Once the cereals had been in use for many centuries, the plants lost the ability to disperse their ripe grains; and harvesting in the later Pre-Pottery Neolithic produced a different kind of gloss on the sickle blades, as the stems of the fully ripe plants were dry and tough. Analysis of the surface of a grinding slab that was found inside one of the small huts produced grains of cereal starch (Nadel et al. 2012; Piperno et al. 2004). The team of researchers have even made a detailed argument based on the identification of some of the seeds as 'weeds' for the beginnings of the practice of cultivation (Snir et al. 2015).

Plant remains may be extremely rare in the context of Epipalaeolithic sites, as explained a few paragraphs ago. There are early Epipalaeolithic sites of a special kind in the Azraq basin, a wetland area in the semi-arid steppe of northeast Jordan, one of which, Kharaneh IV, is currently under investigation. It is an extraordinarily large site covering at least 20,000 m², with an archaeological deposit that is up to 1.5 m thick. The excavators refer to it as an 'aggregation site', where a very large number of early Epipalaeolithic hunter-gatherers came together for considerable periods of time (Maher 2017). Like the occupants of Ohalo II, they made extensive use of the wetland plant species (which are generally unfamiliar to us as plant foods), which have been detected by means of their phytoliths (microscopic silica skeletons of plant parts) in the soil (Ramsey & Rosen 2016); and there were plenty of animals to hunt.

The next sites with plant remains belong late in the late Epipalaeolithic, but there are proxy indicators that suggest the steadily increasing use of hard-seeded grasses and cereals throughout the period. Grinding stones are present in small numbers on sites of late Middle Palaeolithic times onwards (Wright 1994). The numbers and variety of stones for grinding or pounding grows a little in the Upper Palaeolithic, grows more through the Epipalaeolithic period, increasing greatly in the late Epipalaeolithic and more again into the Neolithic. Sickle blades used for harvesting grasses, cereals (but equally possibly for cutting sedges and reeds) are recognisable by the gloss on their cutting edges that results from the friction of the silica-rich stems of grasses, cereals, rushes and the like. Glossed sickle blades have been noted on Epipalaeolithic sites, but they are relatively frequent on late Epipalaeolithic sites. There is no reason to think that people were cutting more and more reeds or rushes (which, in any case, would not have been locally available at a number of sites); it is much more likely that the increasing use of sickles relates to an increasing interest in grasses and cereals. These proxies tell us that in the southern Levant the harvesting of cereals and grasses and their processing through being crushed or ground increased throughout the ENT.

Now we should turn to the north Levant, and three sites in north Syria, two on the banks of the Euphrates, and the third in the gentle hill country to the west. The two Euphrates valley sites were first occupied in the late Epipalaeolithic period. Tell Abu Hureyra was settled shortly before the beginning of the Younger Dryas phase, while Tell Mureybet was first occupied during the Younger Dryas. The landscape around these two sites was semi-arid steppe, but the Euphrates had created a valley whose very narrow floodplain was green and fertile. Both sites were briefly excavated in the early 1970s when a dam was being constructed on the Euphrates that would shortly drown them. The botanists who studied the plant material from Tell Mureybet suggested that the inhabitants may have begun to cultivate the cereals (Van Zeist & Bakker-Heeres 1986), and Sue Colledge, who used multivariate statistical analysis of the plant assemblages at both Tell Mureybet and Abu Hureyra, believed that she could detect weeds of cultivation in association with the cereal grains (Colledge 1998, 2001).

Systematic flotation at Abu Hureyra (probably the first Epipalaeolithic-Neolithic site to be so systematically samples) produced a huge amount of carbonised plant material that has been analysed over decades. The late Gordon Hillman, who was the archaeo-botanist on the team, and a succession of doctoral students and post-doctoral researchers, wrestled for many years with the question of how to interpret the changes in the spectrum of plant remains that could be matched against the onset of the colder, drier conditions of the Younger Dryas phase. In the end, Hillman concluded that, on the balance of probabilities, the slightly increasing size of rye grains with time signalled cultivation on the Euphrates floodplain (Hillman et al. 1989, 2001; Moore et al. 2000). The debate has continued, and Hillman's arguments have been reviewed and refined; Sue Colledge and James Conolly have concluded that the inhabitants certainly sought to broaden the spectrum of small-seeded plants (more work for less returns) in the face of the environmental pressures (Colledge & Conolly 2010). The important argument for the cultivation of rye as the Younger Dryas pressed on the inhabitants of Abu Hureyra had been the recognition of 'weeds of cultivation' alongside the wild rye. Revisiting the analyses, Colledge and Conolly (2010) were reluctant to support the claim that people had begun to cultivate rye. And another group of archaeo-botanists (Weide et al. 2021), reviewing the arguments for early cultivation across the hilly flanks zone, have cast doubt on the ability of botanists to distinguish between plants that were part of the same habitat as wild cereals and 'weeds' that have grown with a cultivated crop in newly disturbed ground; the species that we may class as 'weeds' have to be present among the local flora.

The third late Epipalaeolithic site is Dederiyeh, situated in an ecological zone very different from the two Euphrates valley sites. It lies northwest of Abu Hureyra and Tell Mureybet, about 65 km from the Mediterranean coast in an environment that enjoys 600 mm annual rainfall, and that was open oak woodland in early prehistoric times. The cave site was occupied in the Middle Palaeolithic and Upper Palaeolithic, but late in the late Epipalaeolithic period there was also an occupation in the cave-mouth (Nishiaki et al. 2011). One of the circular houses had been burnt down with its contents intact. The carbonised plant remains include wild einkorn wheat, which was common, some wild barley, small-seeded grasses, especially Stipa (feather grass), a lot of Pistacia, almonds, hackberry and hawthorn. The wood charcoal recovered included pieces of oak and elm, which were the remains of some of the structural beams and posts of the house (Tanno et al. 2013). The chance discovery of an accidentally burnt house in a more generous environment than the semi-arid steppe surrounding the two Euphrates valley sites has produced a quite different picture of the range of plant-food resources accessed by a late Epipalaeolithic community.

Cultivation, Domestication, Garden Agriculture and Farming

What has emerged over recent decades is evidence that the transition from wild cereals to domesticated species was a long process. The common species of wild cereals in Southwest Asia, which can still be found surviving here and there, are two kinds of wheat, einkorn and emmer, and barley. It was these three wild cereals that became the common domesticated crops. It used to be thought that domestication took only a short time (Hillman & Davies 1990). Researchers found that it was necessary to harvest wild wheat or barley before the grains were fully ripe, because, as they became ripe, the plants tend to shed their grains. The rachis is a tiny part that attaches the grain to the stem, and in the wild cereals the rachis becomes brittle as the seed ripens, so that the grains are easily released. However, a very small proportion of wild wheat and barley plants have a mutation that produces a tough rachis that does not become brittle. It was argued that humans harvesting wild stands of einkorn, emmer or barley would act as an unconscious selecting force, as they would tend to gather more of the tough rachis plants and lose some of the brittle rachis plants; since they used seed from their harvest to sow next season's crop, they would help to reproduce the tough rachis mutation. The archaeo-botanist, incidentally, not only needs to be able distinguish between

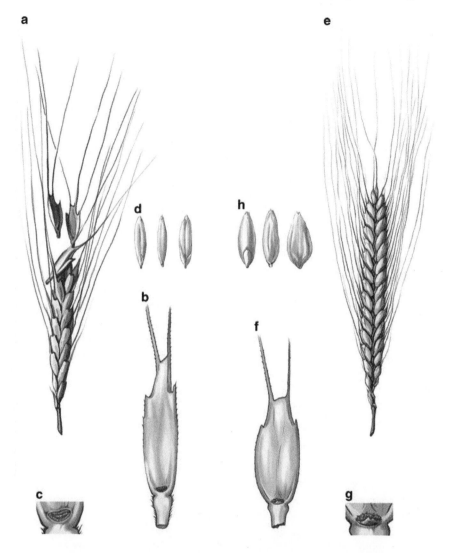

Figure 2.1 Einkorn, (a) an ear of wild einkorn, (e) an ear of domesticated einkorn, (d) wild einkorn grains, (b) wild einkorn spikelet with smooth scar at (c), (f) domesticated spikelet with jagged break at (g), (h) domesticated einkorn grains. (Kilian 2010, Fig. 3. By kind permission of the author, illustration by Svetlana Kilian)

the very similar (carbonised and distorted) grains of the different species, but also to recognise the tiny rachis fragments that are so important. Hillman and Davies (1990) study concluded that the domesticated forms of einkorn, emmer and barley would emerge within 200 years of cultivation, or possibly in only 20 or 30 years, especially if people began to extend the habitat of the wild species by planting crops in fields that they created close to their settlements (Figure 2.1).

Domesticated einkorn, emmer and barley are known in different parts of the hilly flanks of the Fertile Crescent from about 8500 BCE, at or following the beginning of the later Pre-Pottery Neolithic. But, as we have seen, there is accumulating evidence that cultivation was practised at least from the late Epipalaeolithic period, a phase that is labelled pre-domestic cultivation. And we now have direct evidence of the slow transition from the wild species (with small grains and a brittle rachis) to the domesticated species (with large grains and tough rachis). The archaeo-botanist George Willcox was a member of Danielle Stordeur's team that worked on the small, early Pre-Pottery Neolithic settlement of Jerf el Ahmar beside the Euphrates in north Syria. The settlement existed over at least half a millennium, and there were good samples of carbonised plant remains. Willcox was able to measure grain size for both einkorn and barley and plot the average measurements stage by stage through the settlement's long history (Willcox 2004). He found that the size, in particular the fatness, of the grains of both einkorn and barley increased over time (Figure 2.2). The grains at the end of the occupation of the two early Pre-Pottery Neolithic sites were still considerably thinner than the domesticated form from a nearby site dating around 4000 BCE (i.e. a later site within the same environmental zone). Willcox also found evidence of the presence of weeds of cultivation (Willcox et al. 2008). The analysis of the stable isotope of nitrogen in the carbonised cereal grains from Jerf el Ahmar has led to the suggestion that the cultivators were also manuring the soil in which they were cultivating their cereals (Araus et al. 2014). On the other hand, another group have

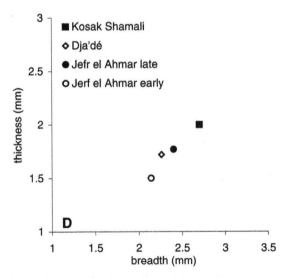

Figure 2.2 Barley grains growing fatter with cultivation over time. The early and later samples from Jerf el Ahmar are complemented by a sample from nearby Dja'dé, which is the same date as the late levels at Jerf el Ahmar. The sample from Kosak Shamali is for reference; the site is also in the Euphrates valley, but later (6th–4th millennium BCE), when fully domesticated. (Willcox 2004, Fig. 3d, with permission)

developed a 'functional ecological model' that set out to distinguish the ecology of arable fields from wild cereal habitats on the basis of how plants respond to mechanical soil disturbance (Weide et al. 2022). They concluded that soil conditions at key pre-domestication cultivation sites such as Jerf el Ahmar were similar to uncultivated wild cereal habitats, implying that pre-domestication cultivation did not create arable environments through regular tillage of the soil. Perhaps sowing the seed by scattering plus simple manuring were enough, over the centuries, to lead to larger-seeded cereals that were better able to take advantage of the artificial conditions. Taken together, the evidence from Jerf el Ahmar and other early Pre-Pottery Neolithic sites demonstrates the intensification of cereal harvesting, storage and processing in the early Pre-Pottery Neolithic (Willcox & Stordeur 2012). There is no evidence that the people of Jerf el Ahmar had any domesticated animals, which leads to the conclusion that they were spreading midden materials from the settlement.

There is evidence for pre-domestic cultivation at early Pre-Pottery Neolithic settlements in the southern Levant (Weiss et al. 2006). In another study, two archaeobotanists collected the data for rachises, brittle or tough (non-shattering), for barley and einkorn from a series of sites between the very beginning of the Pre-Pottery Neolithic around 9700 BCE and the transition between the Pre-Pottery Neolithic and the pottery Neolithic, around or just after 6500 BCE (Asouti & Fuller 2013) (Figure 2.3). Their graphs (cf. Figure 2.4 here) show the slow trend of tough rachis to increase over 3000 years. In another study undertaken with a third colleague, the rates of phenotypic change in crops undergoing domestication were shown to conform to evolutionary theory, but with significant differences with important leguminous crops such as peas and lentils (Fuller et al. 2012).

The idea that Epipalaeolithic 'hunter-gatherers' engaged in cultivation has been difficult for many archaeologists to accept, although geneticists and botanists

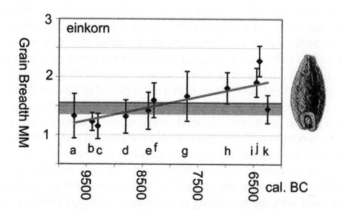

Figure 2.3 Increasing breadth of einkorn grains through time from a number of sites in the hilly flanks zone. Samples from above the horizontal line are the domesticated form. Across the Pre-Pottery Neolithic the wild form gradually increases in breadth. (Fuller et al. 2012, Fig. 2, with permission)

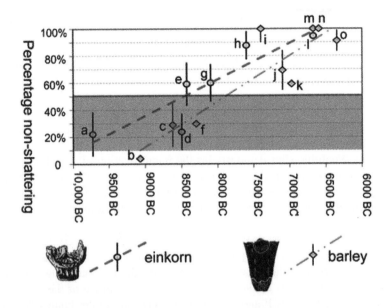

Figure 2.4 Rachis in both einkorn and barley from a series of sites in the hilly flanks zone. The proportion that are tough and do not shatter naturally increases with time, illustrating the slow and gradual process of domestication. (Fuller et al. 2012, Fig. 2, with permission)

have argued that the genetic changes that resulted in domesticated cereals require that their cultivation must have begun three or more thousand years before full domestication can be recognised (Allaby 2010; Allaby et al. 2008; Allaby et al. 2017). A recent microscopic study of the wear-traces on sickle blades, coupled with experimental replication (Ibáñez et al. 2016), showed that there were three classes of wear-traces, of which one was associated with the harvesting of wild cereals and another with domesticated cereals. The third class was associated with the harvesting of cultivated wild cereals at an experimental station in the south of France. In the wild, a head of wheat or barley releases its grains as they ripen from the tip of the head downwards over a short period of time. Anyone harvesting wild cereals, therefore, would try to catch the plants just before they begin to release their grains; in other words, wild cereals would need to be harvested while the stems were relatively green. The cereals have evolved under domestication to retain their grains; domesticated crops are therefore harvested ripe and need to be threshed in order to release the grain. Wild cereals that have begun to be cultivated are somewhere in between. Ibáñez and his colleagues (2016) found that sickle gloss associated with the harvesting of cultivated wild-type cereals occurred at late Epipalaeolithic sites, reinforcing the arguments for the beginnings of cultivation at that time. More controversial is the recent claim that cultivation has been identified 10,000 years earlier, at Ohalo II, at the beginning of the Epipalaeolithic (Snir et al. 2015).

For some time we have concentrated attention on the cereals, but we should not overlook the presence of pulses such as lentils, peas, chickpeas and certain vetches at many or most sites where flotation has produced carbonised wheat or barley parts. Nutritious pulses were widely available around the hilly flanks zone, and in central Anatolia, but it is practically impossible to detect initial cultivation, and the domesticated legumes are little different in size or shape from their wild forms. And they are almost certainly under-represented in archaeological deposits, because, unlike cereals, they do not need a gentle roasting before storage, and are therefore less likely to be accidentally carbonised. Experimental harvesting wild pea, lentil and chickpea in Israel has shown that the labour of gathering the seeds in the wild probably uses more energy than the nutritional value of the small amounts collected per hour (Abbo et al. 2008). The difficulty in harvesting these wild legumes is that they occur widely scattered in small patches (while cereals can appear in dense stands), they are low on the ground, and the pods that contain the seeds are difficult to locate. The conclusion is that the pulses would only be worthwhile if cultivated, which would intensify their density on the ground; the plants could then be harvested whole, allowing the search for the seed-pods to be carried out more efficiently (and less painfully). Since the pulses appear on Epipalaeolithic (and earlier) sites, it is suggested that they may have been cultivated before the pre-domestic cultivation of cereals began; indeed, at some sites pulses have been recovered in greater quantities than cereals (Kislev & Bar-Yosef 1988).

In the late Pre-Pottery Neolithic in the Levant it is obvious to the botanist that the wheat and barley is recognisably domesticated and morphologically different from the wild forms. The grains are much larger, and the rachis – the tiny part that forms the link between the grain and the stem – is found in fragments. Wild cereals shed their seeds when they are ripe; the domesticated cereals have evolved to require human intervention to sow their seeds, and they have a tough rachis that only breaks from the grain when the harvest is threshed. Thus the two ends of the process are recognisable, but how did the cereals progress from their natural state to the domesticated forms. Even when the cereals in the late Pre-Pottery Neolithic are identified by archaeo-botanists as domesticated, they usually represent only a proportion of all the plant remains (Wallace et al. 2018). This is effectively a version of the broad-spectrum strategy; while people engaged in the cultivation of perhaps a couple of cereal species to varying degrees, they continued to rely on a wide variety of wild plant resources, a subsistence strategy that has been called low-level food production (Smith 2001).

Grasses and cereals were of very low significance across the northern arc of the hilly flanks, in the upper Tigris drainage in southeast Turkey, and across northern Iraq in the early Pre-Pottery Neolithic. Hallan Çemi was a small settlement beside a tributary of the Batman Su, itself an important tributary of the Tigris (we saw that its inhabitants hunted a rich variety of large and medium-sized animals, and we shall meet the site again in Chapter 6). It was probably first settled late in the Younger Dryas phase, but its main occupation dates to the early centuries of the early Pre-Pottery Neolithic. The botanical record is diverse, dominated by sea club-rush, followed by knotgrass. Legumes provided less than 10% of the total, and grasses were less than 5% (Savard et al. 2006; Willcox & Savard 2011). The botanical material retrieved from the earliest stages of the settlement of Çayönü, a good deal further upstream in the Tigris drainage, and dating to the later centuries of the early Pre-Pottery Neolithic, included pulses, including bitter vetch, pea, chickpea, and, especially, lentils, which were probably cultivated, sea club-rush, and some wheat and barley (Asouti & Fuller 2013: 324). With the passage of time, the amounts of wheat and barley increase, and in the early centuries of the late Pre-Pottery Neolithic morphologically domesticated cereals begin to appear. At Qermez Dere, on the southern side of the Jebel Sinjar range, overlooking the north Mesopotamian steppe, the botanical assemblage is dominated by small-seeded grasses and large-seeded pulses (vetches, bitter vetch and lentils). Very small amounts of barley and einkorn wheat were found. The assemblage recovered at broadly contemporary M'lefaat, at the edge of the Zagros piedmont, east of Tigris in northeast Iraq, was similar. In general, this arc of early Pre-Pottery Neolithic settlements suggests that late Epipalaeolithic and early Pre-Pottery Neolithic communities were opportunistic in using a broad range of locally available resources and were not particularly focused on grasses and wild cereals.

Körtik Tepe, another settlement in the upper Tigris valley in southeast Anatolia, is particularly interesting. The site was first settled during the Younger Dryas phase and continued through the early Pre-Pottery Neolithic. The study of the wood charcoals shows that the colder, drier climate of the Younger Dryas restricted the growth of the open oak woodland, which probably opened up the environment for more vigorous and varied growth of grasses, cereals and many other seed-producing species (Rössner et al. 2017). The Younger Dryas carbonised plant assemblage is dominated by small-seeded grasses; large-seeded grasses (wild rye, millet, wild barley, and wild einkorn) are present, but in very small quantities. The proportion of cereals and large-seeded grasses increases in the early Holocene/early Pre-Pottery Neolithic settlement and the small-seeded grasses decrease sharply.

Finally, we should turn attention to Gusir Höyük, another settlement in the upper Tigris valley downstream from Körtiktepe, which it resembles in bridging the end of the Younger Dryas phase and the early Pre-Pottery Neolithic. The analysis of the plant remains shows selective use of legumes and nuts during the earlier part of this period, followed by the management of cereal and legume crops from the onset of the early Pre-Pottery Neolithic (Kabukcu et al. 2021). Kabukcu and her collaborators remark that the data from Gusir contrasts with that available from the other upper Tigris valley settlements. They conclude that early Pre-Pottery Neolithic southeast Anatolian plant and animal exploitation strategies were site-specific. It is particularly interesting that they note that this variety is not attributable to environmental constraints, but represents the distinctive identity of each community and its culinary choices.

The piedmont and intermontane valleys of the Zagros region, the eastern wing of the hilly flanks of the Fertile Crescent, where Braidwood began his field research 70 years ago, has for a long time been inaccessible to archaeologists. The early research by Braidwood and several others who followed his lead took place in the early years of radiocarbon dating and was abruptly brought to an end following the

revolution in Iran. The work that had been done did not quite add up to producing a solid cultural sequence; it was just ahead of the implementation of flotation
machines, which means there are poor archaeo-botanical samples; and there were
gaps in the sequence, some rather imprecise and some suspect radiocarbon dates.
Scientific methods, techniques and the development of theory moved ahead based
on fieldwork done elsewhere, and the Zagros region slipped out of sight and out of
mind. The general view grew up that the earliest domesticated plants and animals
were to be found in the Levant; from that perspective, other parts of the hilly flanks
zone, for which there was little useful evidence, were assumed to have followed
from the lead of the Levant, as domesticated plants and animals spread with the
practice of mixed farming. The analysis of materials from the concentration of salvage archaeology sites in the Euphrates valley of north Syria and in southeast Turkey began to subvert that view of the beginnings and spread of agriculture. More
recently, researchers have returned to the Zagros region, both in the Kurdish region
of Iraq and in western and southwestern Iran, and new information has begun to
emerge that supports the idea that the intensification of effort on the use of cereals and the emergence of pre-domestic cultivation occurred in different ways all
around the arc of the hilly flanks zone, including the Zagros region (Willcox 2013).
We now know of settlements dating to the early Pre-Pottery Neolithic period, and
they are producing carbonised cereal remains that show that there was certainly
pre-domestic cultivation of barley, and to a lesser extent wheat, at least as early as
the beginning of the early Pre-Pottery Neolithic (Riehl et al. 2013).

In the early Pre-Pottery Neolithic, then, we see pre-domestic cultivation integrated to varying degrees within an essentially hunter-gatherer subsistence economy (and, as we shall see in Chapter 7, operating within an updated hunter-gatherer
sharing ethos, with communal storage, communal food preparation, and much
communal consumption). Settlements that belong to the very beginning of the late
Pre-Pottery Neolithic are rare, but Tell Aswad, in southern Syria, is informative:
there were quantities of barley rachises of the brittle kind that characterises the
wild species, a confusing amount of domesticated-type chaff. In the late phases of
the site's history, the size of the emmer wheat grains is well within the range of the
domesticated species, there were large lentils indicative of the cultivation of the
domesticated form, and weeds of cultivation have been identified. In the beginning
of the late Pre-Pottery Neolithic, we see the signs of full domestication of cereals,
but it is only in the latter half of that period substantial reliance on a package of
domesticated crops (Kuijt & Goring-Morris 2002). It is surprising that, despite the
number and impressive size of settlements of the late Pre-Pottery Neolithic, there
are few sites whose botanical assemblages have been published; some sites were
excavated before systematic flotation and wet sieving came into use, and some
more recently excavated assemblages are still under intensive study. From the sites
in the southern Levant, each community seems to have preferred to cultivate its
own group of crops, and wild-type cereals and other wild plant foods continued
to be used (Asouti & Fuller 2012: 154). The early eighth millennium BCE levels
at Jericho, one of the first Pre-Pottery Neolithic settlements to be investigated,
produced an unusually full range of cultivated species, including domesticated

barley, emmer and einkorn, flax, lentil and pea. Large settlements of the last half millennium of the late Pre-Pottery Neolithic in Jordan have produced what may be considered a developed 'package' of crops, including emmer wheat, barley, lentils, peas and chickpea. At this stage secondary cereal domesticates, namely free-threshing wheat and naked barley, appear; these are crops that have evolved out of already domesticated species, and they are completely dependent on humans for their reproduction.

Çatalhöyük on the Konya plain in central Anatolia is probably the most thoroughly investigated Neolithic settlement in Southwest Asia. It spans the end of the Pre-Pottery Neolithic into the later, pottery Neolithic (approximately 7400–6000 BCE). House 52 was one of the many houses uncovered during the two and a half decades of recent excavation work, and it gives us a remarkable vignette of both the plant and the animal economy, because the house and its contents were destroyed in a severe conflagration (Twiss et al. 2009). In the late Pre-Pottery Neolithic domestic buildings became more substantial, multi-roomed, and in a number of settlements two storeys high (plus a much-used flat roof). The houses at Çatalhöyük mostly consisted of a large room with a clearly differentiated cooking and food preparation area, and a small inner room for the storage of foodstuffs. In common with houses at many sites of the late Pre-Pottery Neolithic, each house at Çatalhöyük was provided with substantial storage capacity. A clay-built storage bin in the store-room of House 52 contained large amounts of wheat; there were also large amounts of almonds, barley, peas, and a large amount of small-seeded crucifers (which were probably a source of oil).

Once Neolithic communities were engaged in mixed farming the management of their crops, cultivated plots and orchards, and the management of their flocks and herds were closely integrated. The separation of the archaeo-zoological from the archaeo-botanical research into two separate chapters is a necessary artifice. The evidence of stable isotope analysis of carbon (which relates to the availability of water for crop growth) and especially nitrogen (which relates to the manuring of fields) shows that cultivation and herding were integrated and balanced; while people needed feeding, their animals needed foddering, and the yield of crops was greater for the manuring of the fields (Bogaard 2005; Bogaard et al. 2013, 2018).

As with animals, the recently discovered and still incompletely known Pre-Pottery Neolithic of Cyprus has produced surprises for the archaeo-botanists. At Klimonas, a settlement dating to the last centuries of the early Pre-Pottery Neolithic, excavators found pistachio, almond and hackberry (which are commonly found in mainland Levantine early Pre-Pottery Neolithic sites), together with wild barley and emmer wheat (Vigne et al. 2012). Wild cereals are not known in the modern flora of Cyprus, which suggest strongly that they were introduced from the Levantine mainland, where pre-domestic cultivation was practised.

There has been a concentration of fieldwork in the Levant, which has produced much of the raw material for the archaeo-botanists to work with, biasing our view of the whole scene around the hilly flanks of the Fertile Crescent. We have been over-focused on the cereals, which were much in use throughout the Levant (and which are, of course, familiar to us and fundamental to our economies and cultures).

Recent work has shown that we need to expand our view beyond the hilly flanks to include southern central Anatolia and Cyprus. Stepping back to look around the whole of our area, we have learned that there was not a single centre where cultivation lead to the emergence of domesticated forms of cereals and pulses, which were then disseminated all around the hilly flanks. Rather, the picture is a mosaic, with different sub-areas or sites moving along their own pathways with different preferred species, at different paces and times (Arranz-Otaegui et al. 2016). Only at the middle or end of the Pre-Pottery Neolithic period were almost all communities throughout the region growing a more or less standard set of cereals and legumes.

As a final comment on the long process of intensification of the use of plants that led to the domestication of a number of species and ultimately the establishment of agricultural packages, it is important to note that domestication is something important to today's archaeo-botanists; it was long after the time when archaeo-botanists identify the domestication of barley and einkorn and emmer wheat, perhaps at the end of the late Pre-Pottery Neolithic, and more probably in the pottery Neolithic, that farming communities came to rely completely on their farmed crops and herded animals. A recent analysis of the collected wild plant remains from across the whole range of the ENT sought to distinguish those that contributed to the human diet (Wallace et al. 2018). For the Epipalaeolithic period in particular, this study concluded that the very broad spectrum of plants represented in botanical assemblages includes many species that were not collected for food. Having defined a narrower spectrum of collected food plants, the study found that those same wild plant foods were in use both in the early and the late Pre-Pottery Neolithic periods. Even in the final centuries of the late Pre-Pottery Neolithic, people continued to forage in the wild alongside their farming of standard crops.

In both these two chapters that discuss the changes in subsistence strategies for both plant foods and meat, and before we move on to the sequence of chapters describing the archaeology of the ENT, I want to move us away from any idea that the Neolithic can be understood as the conclusion of a series of economic adaptations to or adaptations of environmental resources. Dating from his original articulation of the idea of a broad-spectrum revolution (Flannery 1969), it has been normal to assume that the adaptations in subsistence strategy have been responses to resource depression, whether because of demographic pressure or deteriorating environmental conditions. In her essay reviewing Flannery's broad-spectrum revolution 40 years on, Melinda Zeder pulls together the accumulating arguments against such a simple, one-way process in which the human environment changes and human populations are forced to adapt (Zeder 2012). As we have seen across this and the previous chapter, there is little or no evidence that the two periods of harsher climatic conditions, the Last Glacial Maximum and the Younger Dryas, affected the availability of subsistence resources, except perhaps in the high, intermontane valleys of the Zagros region. And in an earlier chapter we saw that population density increased steadily across the time-range of our period, with the Younger Dryas having no apparent negative effect. As Zeder remarks, especially in the northern arc of the hilly flanks of the Fertile Crescent across southeast Turkey and north Iraq, sites occupied during the Younger Dryas phase seem to have enjoyed abundance of both animal and plant resources. Zeder concludes that niche construction theory, whereby human populations modified their environment to suit their needs, provides a better framework for explaining the changes in subsistence strategies.

Finally, reflecting on the title of this chapter – from foraging to farming – we have a problem in English with the word 'farming'. Although in today's hugely urbanised world hardly any of us have any direct experience of farming, we generally entertain the idea that we know what farming is about. Of course we don't imagine that people in the Neolithic used tractors to pull ploughs and combine harvesters to reap the crops, our image of farming probably includes ploughs drawn by oxen or horses and wagons loaded with the harvest. In English we have no way of distinguishing between what Amy Bogaard and her collaborators call labour-limited and land-limited agriculture (Bogaard et al. 2018, 2019). What we see in the Neolithic is the evolution of labour-limited agriculture. Bogaard alternatively refers to labour-limited agriculture as 'garden-farming', where production is limited by the physical labour, which may involve manuring as well as digging and sowing, by whoever is doing the cultivation by hand. There are four factors to land-limited agriculture; labour, manure, land (as in labour-limited agriculture), and animal traction. A single farmer with an ox team and a plough could cultivate as much land as would require between 2 and 15 people cultivating by hand (Halstead 1995, 2014). And, with a wagon or a sledge, the harvest could be hauled home. Bogaard et al.'s research was concerned with the emergence of wealth inequalities, and they conclude that land-ownership plus ownership of a team or teams of oxen provided the basis for the inter-generational accumulation of material wealth. Since it is often assumed that the emergence of farming produced inequalities of wealth for the first time (leading to competition, conflict and even warfare), it is important here to assert that the simple 'labour-limited' agriculture that emerged over the Neolithic is very different from the 'land-limited' farming that only emerged some time later.

References

Abbo, S., Zezak, I., Schwartz, E., Lev-Yadun, S., Kerema, Z., & Gopher, A. (2008). Wild Lentil and Chickpea Harvest in Israel: Bearing on the Origins of Near Eastern Farming. *Journal of Archaeological Science*, 35(12), 3172–7.

Allaby, R. G. (2010). Integrating the Processes in the Evolutionary System of Domestication. *Journal of Experimental Botany*, 61(4), 935–44.

Allaby, R. G., Fuller, D. Q., & Brown, T. A. (2008). The Genetic Expectations of a Protracted Model for the Origins of Domesticated Crops. *Proceedings of the National Academy of Sciences of the United States of America*, 105(37), 13982–6.

Allaby, R. G., Stevens, C., Lucas, L., Maeda, O., & Fuller, D. Q. (2017). Geographic Mosaics and Changing Rates of Cereal Domestication. *Philosophical Transactions of the Royal Society B-Biological Sciences*, 372(1735). doi:10.1098/rstb.2016.0429

Araus, J. L., Ferrio, J. P., Voltas, J., Aguilera, M., & Buxó, R. (2014). Agronomic Conditions and Crop Evolution in Ancient Near East Agriculture. *Nature Communications*, 5(1), 3953.

Arranz-Otaegui, A., Colledge, S., Zapata, L., Teira-Mayolini, L. C., & Ibáñez, J. J. (2016). Regional Diversity on the Timing for the Initial Appearance of Cereal Cultivation and Domestication in Southwest Asia. *Proceedings of the National Academy of Sciences,* 113(49), 14001–6. doi:10.1073/pnas.1612797113

Asouti, E., & Fuller, D. (2012). From Foraging to Farming in the Southern Levant: The Development of Epipalaeolithic and Pre-Pottery Neolithic Plant Management Strategies. *Vegetation History and Archaeobotany,* 21(2), 149–62.

Asouti, E., & Fuller, D. (2013). A Contextual, Practice-Based Approach to the Origins of Agriculture. *Current Anthropology,* 54(3), 299–345.

Binford, L. R. (1968). Post-Pleistocene adaptations. In L. R. Binford & S. R. Binford (Eds.), *New Perspectives in Archaeology* (pp. 313–341). Chicago: University of Chicago Press.

Binford, L. R. (1980). Willow Smoke and Dogs Tails: Hunter-Gatherer Settlement Systems and Archeological Site Information. *American Antiquity,* 45, 4–20.

Bogaard, A. (2005). 'Garden Agriculture' and the Nature of Early Farming in Europe and the Near East. *World Archaeology,* 37(2), 177–96.

Bogaard, A., Fochesato, M., & Bowles, S. (2019). The Farming-Inequality Nexus: New Insights from Ancient Western Eurasia. *Antiquity,* 93(371), 1129–43. doi:10.15184/aqy.2019.105

Bogaard, A., Fraser, R., Heaton, T. H. E., Wallace, M., Vaiglova, P., Charles, M., Jones, G., Evershed, R. P., Styring, A. K., Andersen, N. H., Arbogast, R.-M., Bartosiewicz, L., Gardeisen, A., Kanstrup, M., Maier, U., Marinova, E., Ninov, L., Schäfer, M., & Stephan, E. (2013). Crop Manuring and Intensive Land Management by Europe's First Farmers. *Proceedings of the National Academy of Sciences,* 110(31), 12589–94.

Bogaard, A., Styring, A., Whitlam, J., Fochesato, M., Bowles, S. (2018). Farming, Inequality and Urbanization: A Comparative Analysis of Late Prehistoric Northern Mesopotamia and South-West Germany. In T. A. Kohler & M. E. Smith (Eds.), *Ten Thousand Years of Inequality: The Archaeology of Wealth Differences* (pp. 201–29). Tucson: University of Arizona Press.

Childe, V. G. (1936). *Man Makes Himself.* London: C. A. Watts.

Colledge, S. (1998). Identifying Pre-Domestication Cultivation Using Multivariate Analysis. In A. B. Damania, J. Valkoun, G. Willcox & C. O. Qualset Aleppo (Eds.), *The Origins of Agriculture and Crop Domestication: The Harlan Symposium* (pp. 121–30). Syria: GRCP, ICARDA, IPGRI, FAO & Univ of California.

Colledge, S. (2001). *Plant Exploitation on Epipalaeolithic and Early Neolithic Sites in the Levant,* Oxford: British Archaeological Reports, International Series 986.

Colledge, S., & Conolly, J. (2010). Reassessing the Evidence for the Cultivation of Wild Crops during the Younger Dryas at Tell Abu Hureyra, Syria. *Environmental Archaeology,* 15(2), 124–38.

Flannery, K. V. (1969). The origins and ecological effects of early domestication in Iran and the Near East. In P. J. Ucko & G. W. Dimbleby (Eds.), *The Domestication and Exploitation of Plants and Animals* (pp. 73–100). London: Duckworth.

Fuller, D., Asouti, E., & Purugganan, M. (2012). Cultivation as Slow Evolutionary Entanglement: Comparative Data on Rate and Sequence of Domestication. *Vegetation History and Archaeobotany,* 21(2), 131–45.

Groman-Yaroslavski, I., Weiss, E., & Nadel, D. (2016). Composite Sickles and Cereal Harvesting Methods at 23,000-Years-Old Ohalo II, Israel. *PLoS ONE,* 11(11), e0167151. doi:10.1371%2Fjournal.pone.0167151

Halstead, P. (1995). Plough and Power: The Economic and Social Significance of Cultivation with the Ox-Drawn Ard in the Mediterranean. *Bulletin on Sumerian Agriculture*, 8, 11–22.

Halstead, P. (2014). *Two Oxen Ahead: Pre-Mechanized Farming in the Mediterranean.* Cambridge: Wiley Blackwell.

Harlan, J. R., & Zohary, D. (1966). Distribution of Wild Wheats and Barley. *Science*, 153(3740), 1074–80.

Hillman, G. C., Colledge, S. M., & Harris, D. R. (1989). Plant-Food Economy during the Epipalaeolithic Period at Tell Abu Hureyra, Syria: Dietary Diversity, Seasonality, and Modes of Exploitation. In D. R. Harris & G. C. Hillman (Eds.), *Foraging and Farming* (pp. 240–68). London: Unwin Hyman.

Hillman, G. C., & Davies, M. S. (1990). Measured Domestication Rates in Wild Wheats and Barley under Primitive Cultivation, and Their Archaeological Implications. *Journal of World Prehistory*, 4(2), 157–222.

Hillman, G. C., Hedges, R., Moore, A. M. T., Colledge, S., & Pettitt, P. (2001). New Evidence of Lateglacial Cereal Cultivation at Abu Hureyra on the Euphrates. *The Holocene*, 11(4), 383–93.

Hole, F., & Flannery, K. (1968). The Prehistory of Southwestern Iran: A Preliminary Report. *Proceedings of the Prehistoric Society,* 33, 147–206.

Hole, F., Flannery, K. V., & Neely, J. A. (1969). *Prehistory and Human Ecology of the Deh Luran Plain; An Early Village Sequence from Khuzistan, Iran.* Memoirs of the Museum of Anthropology, University of Michigan, no. 1. Ann Arbor: Museum of Anthropology, University of Michigan.

Ibáñez, J. J., Anderson, P. C., González-Urquijo, J., & Gibaja, J. (2016). Cereal Cultivation and Domestication as Shown by Microtexture Analysis of Sickle Gloss through Confocal Microscopy. *Journal of Archaeological Science*, 73, 62–81.

Kabukcu, C., Asouti, E., Pöllath, N., Peters, J., & Karul, N. (2021). Pathways to Plant Domestication in Southeast Anatolia Based on New Data from Aceramic Neolithic Gusir Höyük. *Scientific Reports*, 11(1), 2112. doi:10.1038/s41598-021-81757-9

Kislev, M. E., & Bar-Yosef, O. (1988). The Legumes: The Earliest Domesticated Plants in the Near East. *Current Anthropology*, 29(1), 175–9.

Kislev, M. E., Nadel, D., & Carmi, I. (1992). Epipalaeolithic (19,000 BP) Cereal and Fruit Diet at Ohalo-II, Sea of Galilee, Israel. *Review of Palaeobotany and Palynology*, 73(1–4), 161–6.

Kuijt, I., & Goring-Morris, A. N. (2002). Foraging, Farming, and Social Complexity in the Pre-Pottery Neolithic of the Southern Levant: A Review and Synthesis. *Journal of World Prehistory*, 16(4), 361–440.

Lee, R. B. (1968). What Hunters Do for a Living, or How to Make Out on Scarce Resources. In R. B. Lee & I. DeVore (Eds.), *Man the Hunter* (pp. 30–48). Chicago, IL: Aldine

Lee, R. B., & DeVore, I. (1968). *Man the Hunter.* Chicago, IL: Aldine.

Maher, L., (2017). Late Quaternary Refugia, Aggregations, and Palaeoenvironments in the Azraq Basin, Jordan. In Y. Enzel & O. Bar-Yosef (Eds.), *Quaternary of the Levant: Environments, Climate Change, and Humans* (pp. 679–90). Cambridge: Cambridge University Press.

Moore, A. M. T., Hillman, G. C., & Legge, A. J. (2000). *Village on the Euphrates: From Foraging to Farming at Abu Hureyra*, Oxford: Oxford University Press.

Nadel, D. (2017). Ohalo II: A 23,000-Year-Old Fisher-Hunter-Gatherer's Camp on the Shore of Fluctuating Lake Kinneret (Sea of Galilee). In O. Bar-Yosef & Y. Enzel (Eds.),

Quaternary Environments, Climate Change, and Humans in the Levant (pp. 291–4). Cambridge: Cambridge University Press.

Nadel, D., Piperno, D. R., Holst, I., Snir, A., & Weiss, E. (2012). New Evidence for the Processing of Wild Cereal Grains at Ohalo II, a 23,000-Year-Old Campsite on the Shore of the Sea of Galilee, Israel. *Antiquity*, 86(334), 990–1003.

Nadel, D., & Werker, E. (1999). The Oldest Ever Brush Hut Plant Remains from Ohalo II, Jordan Valley, Israel (19,000 BP). *Antiquity*, 73, 755–64.

Nishiaki, Y., Muhesen, S., & Akazawa, T. (2011). Newly Discovered Late Epipalaeolithic Assemblages from Dederiyeh Cave, the Northern Levant. In E. Healey, S. Campbell, & O. Maeda (Eds.), *Studies in Technology, Environment, Production, and Society. Proceedings of the 6th International Conference on PPN Chipped and Ground Stone Industries of the Fertile Crescent (Manchester)* (pp. 79–87). Berlin: Ex Oriente.

Piperno, D. R., Weiss, E., Holst, I., & Nadel, D. (2004). Starch Grains on a Ground Stone Implement Document Upper Paleolithic Wild Cereal Processing at Ohalo II, Israel. *Nature*, 430, 670–3. doi:10.1038/nature02734

Ramsey, M. N., & Rosen, A. M. (2016). Wedded to Wetlands: Exploring Late Pleistocene Plant-Use in the Eastern Levant. *Quaternary International*, 396, 5–19.

Riehl, S., Zeidi, M., & Conard, N. J. (2013). Emergence of Agriculture in the Foothills of the Zagros Mountains of Iran. *Science*, 341(6141), 65–7.

Rössner, C., Deckers, K., Benz, M., Özkaya, V., & Riehl, S. (2017). Subsistence Strategies and Vegetation Development at Aceramic Neolithic Körtik Tepe, Southeastern Anatolia, Turkey. *Vegetation History and Archaeobotany*, 27, 15–29.

Sahlins, M. (1968). Notes on the Original Affluent Society. In R. B. Lee & I. DeVore (Eds.), *Man the Hunter* (pp. 85–9). New York, NY: Aldine.

Sahlins, M. (1972). *Stone Age Economics*. Chicago, IL: Aldine-Atherton.

Savard, M., Nesbitt, M., & Jones, M. (2006). The Role of Wild Grasses in Subsistence and Sedentism: New Evidence from the Northern Fertile Crescent. *World Archaeology*, 38(2), 179–96.

Smith, B. D. (1994). *The Emergence of Agriculture*, New York, NY: Scientific American Library.

Smith, B. D. (2001). Low-Level Food Production. *Journal of Archaeological Research*, 9(1), 1–43.

Snir, A., Nadel, D., Groman-Yaroslavski, I., Melamed, Y., Sternberg, M., Bar-Yosef, O., & Weiss, E. (2015). The Origin of Cultivation and Proto-Weeds, Long before Neolithic Farming. *PLoS ONE*, 10(7), e0131422.

Tanno, K., Willcox, G., Muheisen, S., Nishiaki, Y., Kanjo, Y., & Akazawa, T. (2013). Preliminary Results from the Analyses of Charred Plant Remains from a Burnt Natufian Building at Dederiyeh Cave in Northwest Syria. In O. Bar-Yosef & F. o. R. Valla (Eds.), *Natufian Foragers in the Levant: Terminal Pleistocene Social Changes in Western Asia* (pp. 83–7). Ann Arbor, MI: International Monographs in Prehistory.

Testart, A. (1982). The Significance of Food Storage among Hunter-Gatherers: Residence Patterns, Population Densities and Social Inequalities. *Current Anthropology*, 23, 523–37.

Twiss, K. C., Bogaard, A., Charles, M., Henecke, J., Russell, N., Martin, L., & Jones, G. (2009). Plants and Animals Together: Interpreting Organic Remains from Building 52 at Çatalhöyük. *Current Anthropology*, 50(6), 885–95.

Van Zeist, W., & Bakker-Heeres, J. A. H. (1986). Archaeobotanical Studies in the Levant 3. Late-Palaeolithic Mureybet. *Palaeohistoria*, 26, 171–99.

Vigne, J.-D., Briois, F., Zazzo, A., Willcox, G., Cucchi, T., Thiébault, S., Carrèe, I., Franel, Y., Touquet, R., Martin, C., Moreau, C., Comby, C., & Guilaine, J. (2012). First Wave of Cultivators Spread to Cyprus at Least 10,600 y Ago. *Proceedings of the National Academy of Sciences*, 109(22), 8445–9.

Wallace, M., Jones, G., Charles, M., Forster, E., Stillman, E., Bonhomme, V., Livarda, A., Osborne, C. P., Rees, M., Frenck, G., & Preece, C. (2019). Re-Analysis of Archaeobotanical Remains from Pre- and Early Agricultural Sites Provides No Evidence for a Narrowing of the Wild Plant Food Spectrum during the Origins of Agriculture in Southwest Asia. *Vegetation History and Archaeobotany* 28, 449–463.

Weide, A., Green, L., Hodgson, J. G., Douché, C., Tengberg, M., Whitlam, J., . . . Bogaard, A. (2022). A New Functional Ecological Model Reveals the Nature of Early Plant Management in Southwest Asia. *Nature Plants*. doi:10.1038/s41477-022-01161-7

Weide, A., Hodgson, J. G., Leschner, H., Dovrat, G., Whitlam, J., Manela, N., Melamed, Y. O., & Bogaard, A. (2021). The Association of Arable Weeds with Modern Wild Cereal Habitats: Implications for Reconstructing the Origins of Plant Cultivation in the Levant. *Environmental Archaeology*, 1–16. doi:10.1080/14614103.2021.1882715

Weiss, E., Kislev, M. E., & Hartmann, A. (2006). Autonomous Cultivation before Domestication. *Science*, 312(5780), 1608–10.

Weiss, E., Kislev, M. E., Simchoni, O., Nadel, D., & Tschauner, H. (2008). Plant-Food Preparation Area on an Upper Paleolithic Brush Hut Floor at Ohalo II, Israel. *Journal of Archaeological Science*, 35(8), 2400–14.

Weiss, E., Wetterstrom, W., Nadel, D., & Bar-Yosef, O. (2004). The Broad Spectrum Revisited: Evidence from Plant Remains. *Proceedings of the National Academy of Sciences of the United States of America,* 101(26), 9551–5.

Willcox, G. (2004). Measuring Grain Size and Identifying Near Eastern Cereal Domestication: Evidence from the Euphrates Valley. *Journal of Archaeological Science*, 31, 145–50.

Willcox, G. (2013). The Roots of Cultivation in Southwestern Asia. *Science*, 341(6141), 39–40.

Willcox, G., Fornite, S., & Herveux, L. (2008). Early Holocene Cultivation before Domestication in Northern Syria. *Vegetation History and Archaeobotany*, 17(3), 313–25.

Willcox, G., & Savard, M. (2011). Botanical Evidence for the Adoption of Cultivation in Southeast Turkey. In M. Özdogan, N. Başgelen, & P. Kuniholm (Eds.), *The Neolithic in Turkey, New Excavations and New Research* (pp. 267–80). Istanbul: Archaeology and Art Publications.

Willcox, G., & Stordeur, D. (2012). Large-Scale Cereal Processing before Domestication during the Tenth Millennium cal BC in Northern Syria. *Antiquity*, 86(331), 99–114.

Woodburn, J. (1982). Egalitarian Societies. *Man*, 17, 431–45.

Wright, K. I. (1994). Ground Stone Tools and Hunter–Gatherer Subsistence in Southwest Asia: Implications for the Transition to Farming. *American Antiquity*, 59(2), 238–63.

Zeder, M. A. (2012). The Broad Spectrum Revolution at 40: Resource Diversity, Intensification, and an Alternative to Optimal Foraging Explanations. *Journal of Anthropological Archaeology*, 31(3), 241–64.

3   Changing Subsistence Strategies
Hunting and Herding

This chapter forms a pair with the previous chapter, looking at the present state of our knowledge concerning the changes in strategies of subsistence. While there is evidence from both plant and faunal remains, animal bones survive in the soil, especially in the limestone derived soils of much of southwest Asia, while traces of vegetable food sources are rare, and biased when we have to rely on accidentally carbonised plant remains. It is therefore obvious, but it has to be said, that we cannot assess the relative importance of plant foods and meat in the diet of any community. This chapter is concerned with changes in hunting strategies, the beginnings of animal management, the identification of the domestication of sheep, goat, cattle and pig, and the development of herding strategies. Following from the early part of the preceding chapter, the Epipalaeolithic groups who we shall be meeting in the next two chapters were of great importance to our story. Braidwood had hypothesised that (Epipalaeolithic) hunter-gatherers would have intensified plant-food collection, leading to incipient cultivation; Flannery (1969) sought to document his broad-spectrum revolution theory in what was known of the Epipalaeolithic at that time; like Binford (1968), he proposed that, as hunter-gatherer peoples began to reduce mobility and to settle, their rate of population growth expanded, and increasing human population density made it necessary for them to extend the range of animals, birds and reptiles that they hunted. But that increased reliance on smaller prey only increased the rate of population growth, requiring further intensification of their efforts to supply their subsistence needs; and that intensification led to the beginnings of cultivation and herding of animals.

The Changing Spectrum: Fauna

Changes through time in the spectrum of prey species were noted 40 years ago by Simon Davis (1983). Working with data from sites in the southern Levant he remarked a steady decline in the proportions of larger herd animals and increasing reliance on, and then pressure on, gazelle populations. He looked at the statistics for a sequence of bone assemblages, starting at the end of the Middle Palaeolithic, and proceeding through Upper Palaeolithic and Epipalaeolithic periods. He returned to the subject, examining data from a cluster of sites that were all from a single environmental zone in northern Israel (Davis 2005). The Middle Palaeolithic baseline

DOI: 10.4324/9781351069281-4

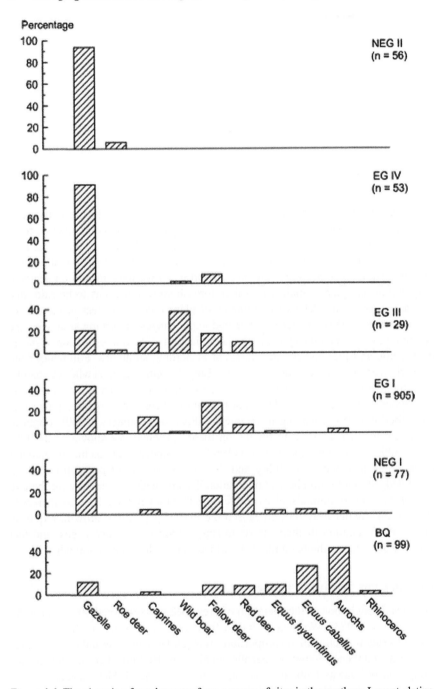

Figure 3.1 The changing faunal spectra from a group of sites in the southern Levant, dating from the late Middle Palaeolithic (at the bottom) through the Upper Palaeolithic and Epipalaeolithic. (Davis 2005, Fig.2, with kind permission of the author)

spectrum consisted of mostly large species, particularly aurochs, wild equids, red deer and even some rhinoceros. Through the Upper Palaeolithic and Epipalaeolithic the spectrum shifted. First, the large species almost disappear, and red deer, then (the smaller) fallow deer, and wild boar became the mainstay of the hunters. But, as the Epipalaeolithic period progressed, gazelle became by far the dominant species providing meat. The significance of these shifts in the spectrum is easily appreciated when one reflects that the meatweight of an aurochs was between 400 and 800 kg (cows were a good deal smaller than bulls), deer ranged between 50 and 200 kg, while gazelle were between 20 and 50 kg (Stutz et al. 2009: Figure 2) (Figure 3.1).

We can take the story forward with the bone assemblage from the site of Hatoula in Israel which was analysed by a team of three specialists (Davis et al. 1994). Davis worked on the mammals and his colleagues studied the bird and the fish bones. The settlement was occupied at the end of the Epipalaeolithic and again at the beginning of the Neolithic. The trio noted that there were increasing efforts through time to increase the amounts of both fish and birds. At the same time increased reliance on gazelle reached a stage in the earliest Neolithic where the extraordinarily high proportions of juvenile gazelle indicate that the species was being over-hunted. What is remarkable about the relatively large quantities of fishbone is that they were predominantly marine species, while Hatoula, on the west side of the Judaean hills, was 28 km from the sea. One interpretation of these trends would be that it represents a long-term depression in the availability of resources; hunters needed to spend more time and effort in hunting for smaller returns. Those who adopt this interpretation would point to the subsequent domestication of sheep and goat as a means of resolving the increasingly serious problem or resource availability. There is, however, another and rather different interpretation, which we will come to later (Figure 3.2).

Closer study of the faunal assemblages from Epipalaeolithic sites in the southern Levant has shown that the declining proportion of bones of large and middle-sized prey indeed corresponds to rising numbers of bones of small mammals, birds and reptiles, particularly tortoises (M. Stiner & Munro 2002). Within these data there is a significant trend among the small species. In the early Epipalaeolithic the tortoises that were taken were full-size adults, but the later tortoises were rather smaller, immature specimens, and in the late Epipalaeolithic tortoise numbers decline steeply. To compensate for the decline of tortoises the numbers of small mammals such, as hares and foxes, and birds rise sharply (Stutz et al. 2009). Tortoises are slow-moving and easily caught, but they are also slow in reproducing and maturing, which makes them vulnerable to predation in the presence of the sedentary hunter-gatherers of the late Epipalaeolithic (which is what we will be finding in the next chapter). The small mammals and birds are fast-moving and more challenging to hunt or trap, but they reproduce quickly and can withstand the predation of sedentary hunter-gatherers (Figure 3.3).

Our information for the Epipalaeolithic period is strongly biased by the great number of sites of that period in the southern Levant, and the scarcity of excavated sites whose assemblages have been closely studied elsewhere in the hilly flanks

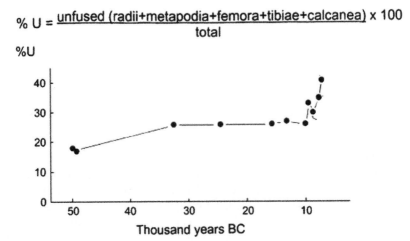

$$\% \text{ U} = \frac{\text{unfused (radii+metapodia+femora+tibiae+calcanea)}}{\text{total}} \times 100$$

Figure 3.2 Percentages of immature gazelle from a cluster of sites in the north of Israel from the Middle Palaeolithic to the early Pre-Pottery Neolithic, when the percentages increase greatly. Davis interprets this rapid increase as evidence for over-hunting, which was immediately followed by a switch to the herding of sheep and goat. (Davis 2005, Fig.4, with kind permission of the author)

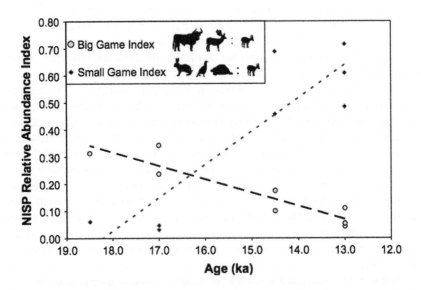

Figure 3.3 Declining abundance of larger species versus increased numbers of smaller species from a series of Epipalaeolithic sites in the southern Levant, illustrating the trend to broad-spectrum strategies in the region. (Stutz et al. 2009, Fig. 4, with permission)

of the Fertile Crescent. It would be wrong to generalise for the whole of the hilly flanks zone from the relatively high-quality data from the southern Levant. The floral and faunal material from the site of Hallan Çemi in southeast Turkey shines a light in an otherwise still poorly documented region; it contrasts with the data from the southern Levant. Hallan Çemi was a small settlement in the upper Tigris drainage in southeast Anatolia (we shall encounter it again in the next chapter). It was occupied at the end of the Younger Dryas phase and into the earliest centuries of the Neolithic period (Rosenberg 2011). In several ways Hallan Çemi requires the re-thinking of old assumptions. The hunters of Hallan Çemi brought home plentiful meat that was made up of sheep, goat, red deer and wild pig; amounts of small mammals, birds, and tortoise were very low, and the tortoise were mainly full-size adults (Starkovich & Stiner 2009). Even at the end of the Younger Dryas phase in a valley deep in the Taurus foothills, the territory around their settlement offered them a plentiful supply of medium-sized animals. The two tons of recovered animal bones make a reliable sample and they are well-studied. Her work on a portion of the material has convinced Melinda Zeder that the people of Hallan Çemi 'sustained a subsistence economy based on a wide range of plentiful and predictably available resources that were capable of supporting a sedentary, stable community for several hundred years' (Zeder 2012: 248). Hallan Çemi contradicts the idea that cultivation of crops and herding of animals was a (Neolithic) response to resource depression, amplified by the effects of the Younger Dryas arid, cold phase.

As is indicated by the hunted wild fauna at Hallan Çemi, all around the arc of the hilly flanks at the very start of the early Pre-Pottery Neolithic, settlements of this first part of the Pre-Pottery Neolithic continued to rely on hunting. And, as in the Epipalaeolithic, the species on which each community relied reflected the local environment. In north Iraq, the very early Pre-Pottery Neolithic sites of Qermez Dere and M'lefaat produced mainly gazelle, sheep and goat bones, with smaller numbers of hare and fox, and occasional equids and wild cattle bones (Watkins et al. 1995: 13). At Tell Mureybet and Abu Hureyra, settlements beside the Euphrates in north Syria where occupation began at the time of the Younger Dryas cold phase, equids (an extinct wild ass species), gazelle, and some wild cattle (aurochs) were the main prey species, supported by amounts of fox, hare, and migrant bird species (Helmer et al. 2004: 146).

When did things change? When and where did people first come to rely on herding domesticated animals? How did that happen? At the end of the Pleistocene around the hilly flanks zone there was a rich spectrum of large and medium-sized herd animals, and a variety of small mammals, many birds, and, in some localities, amphibians, reptiles and fish. Among them there were sheep and goat, cattle and pig, the four species that were domesticated at some time during the Neolithic, and became the basis of farming economies. The recognition of domestication has proved problematic. Kathleen Kenyon's colleague at the London Institute of Archaeology, Frederick Zeuner, analysed the animal bones from her 1950s excavations at Pre-Pottery Neolithic Jericho. He proposed that the goat was domesticated on the basis of morphological changes; he noted that their horns were more twisted than is the case with examples in reference collections of wild goat bones (Zeuner

1955). But other zoologists were sceptical, noting that there were minor morphological variations among different populations of wild goat, and that the Jericho goats could equally be a local wild stock. Similarly, when it was noted that (domesticated?) sheep and goat from Neolithic settlements were smaller than the wild animals from Palaeolithic sites, zoologists responded that a number of wild species had become smaller in the warmer, more stable Holocene environment.

Archaeo-zoologists turned to the age and sex profiles of bone assemblages. At its simplest, hunters would kill whatever animal in a group of gazelle or wild sheep that they could, and the age and sex distribution of the archaeological bone assemblage for that species would approximately reflect the mortality pattern of the wild species. Farmers with a flock of (domesticated) sheep would cull the young males ahead of winter, because of the costs of over-wintering them versus the relatively small gain in meatweight if they were kept. Keeping the females for future breeding was an obvious strategy, and a skewed age and sex distribution pattern would be present in the bones, with many young (male) animals and the mature animals being predominantly female. But what if the wild males and females of a species tended to live in separate groups for most of the year, and hunters focused their predation on the flocks of females with young? (Figure 3.4).

The recognition of the domestication of a species has proved difficult; it requires large assemblages of animal bone (the majority of recovered pieces of animal bone are too fragmentary to be diagnostic), and very careful study. Sheep and goat are particularly difficult to differentiate from the skeletal material; only a few distinctive characteristics are present on certain bones. As a non-expert in archaeozoology, I rely in the following paragraphs on the views of leading specialists.

Figure 3.4 A typical wild sheep. Sheep and goat can be identified and differentiated by only a few bones. The wild and early domesticated sheep had coats of hair like goats; the fleece with which we are familiar was bred into sheep somewhat later.

Something on which they agree is that domestication was a process that extended over a lengthy period, the early stages of which are hard to identify.

Dogs were domesticated in the late Epipalaeolithic in Southwest Asia (and considerably earlier in Europe). They may have been used in support of hunting, and the safest identification of their presence is when dogs have been found buried together with a human body (Davis & Valla 1978; Tchernov & Valla 1997; Yeomans et al. 2019). Our interest, however, mainly concerns the exploitation of animals for meat and other products. Sheep and goat were the first species to be taken under human control. The claim that wild sheep and goat were being managed by the hunter-gatherer group occupying the small settlement of Zawi Chemi in a valley of the Zagros in remote northeastern Iraq (Perkins 1964) was for a long time thought to be out on a limb and somewhat improbable. The site was rather imprecisely radiocarbon dated to the boundary between the Epipalaeolithic and the Neolithic, which seemed to be much earlier than expected for the herding of domesticated sheep or goat, and the sheep and goat bones belonged to wild populations. However, Perkins noted that there were very high proportions of young animals, which seemed to indicate that people were able to cull selectively in a way that managed the flock sustainably. A recent re-analysis of the bones has confirmed Perkins' work and supports the conclusion that local populations of wild sheep and goat were being managed in some way (Zeder 2011). Zeder has noted that Zawi Chemi now falls into a pattern of herd management of sheep and goat that can be recognised at contemporary settlements in the upper Tigris basin in southeast Turkey (Zeder 2012).

Long-lived settlement sites can sometimes give us a view of changing practices over several or many centuries. The settlement of Çayönü in the upper Tigris drainage in southeast Turkey was long-lived through the Pre-Pottery Neolithic period, and the faunal remains from successive strata show gradual reduction in size and morphology of sheep, goat and pig (Hongo et al. 2002, 2004, 2005). In the earliest stages the sheep, goat, pig and cattle appear typical of the wild forms, but the initial stages of management could be detected in the appearance of some small-sized individuals and subtle changes in the kill-off patterns, as well as in the changes in stable isotope ratios of carbon and nitrogen, which reflect a change in their diet, arguably because of restricted pasturing and winter foddering. There was an overall decrease in the proportion of wild species in the assemblages, before marked size reduction and an increase in the proportion of females in the assemblage appeared in the mid- to late stages of the later Pre-Pottery Neolithic. By the closing stages of the settlement's history, the population depended heavily on their fully domesticated sheep and goat.

The archaeo-zoologist who analysed the faunal assemblage from the site of Tell Aswad, a later Pre-Pottery Neolithic settlement (from 8500 BCE) not far from Damascus in southern Syria (Ducos 1993), noted that the goats on which the population depended in the early stages of the settlement were morphologically wild, but that wild goat would not be found in the environment of the Damascus basin, at a significant distance from the mountains on the Syria-Lebanon border. In the following centuries, the goat bones showed that young male animals were being selectively culled, and the adult animals were beginning to change towards a typical domesticated population.

The earliest settlement site that Hole and Flannery found and excavated on the Deh Luran plain of southwest Iran, close to the Iraq-Iran frontier, produced large quantities of goat bones. Both the seed for planting their cereal crops and the goats on which they depended would not have been available in the low-lying alluvial environment, and Flannery concluded that the founders of the settlement of Ali Kosh brought with them the resources that they were accustomed to using in the Zagros valleys where they were native (Hole et al. 1969; Hole & Flannery 1968).

A recent study of the genomes of early domesticated goat from around the hilly flanks zone demonstrates something important that is probably true of the domestication of the other species, sheep, cattle and pig, and has been shown to be true of the domestication of the cereals wheat and barley (Daly et al. 2018). Domesticated goats appear at almost the same time all around the hilly flanks zone, and the genetic evidence shows that multiple wild populations contributed to the early domesticated goat populations; goat domestication was a mosaic of local or regional domestications.

Turning south to the Euphrates valley in north Syria, at the final Epipalaeolithic and Pre-Pottery Neolithic settlements of Tell Mureybet and Abu Hureyra, gazelle and wild equids were the main prey of hunters in the earliest stages, and sheep continued to be only a minority (5%) of the animals hunted in the early levels of the Pre-Pottery Neolithic of Abu Hureyra. However, a recently published paper (Smith et al. 2022) reports that dung spherulites had been found, preserved within flotation samples from both the final Epipalaeolithic and the later Pre-Pottery Neolithic occupations, implying that sheep or goat were kept among the buildings of the settlement. Some darkened spherulites were identified within an Epipalaeolithic period firepit, indicating the use of dung as fuel. When the site was reoccupied at the beginning of the later Pre-Pottery Neolithic, the substantial rectilinear buildings were quite closely packed, and the concentrations of dung spherulites were much reduced. At a certain moment in the later Pre-Pottery Neolithic occupation the situation suddenly changed, and domesticated sheep came to represent more than 50% of the animal bones (Legge & Rowley-Conwy 2000). It appears that, in that part of the Euphrates valley, although some sheep were being managed in the settlement in the final Epipalaeolithic occupation, the appearance of morphologically domesticated animals is the result of the arrival of already domesticated sheep from elsewhere. The sudden appearance of domesticated sheep (introduced from further north?) was similarly noted at the nearby settlement of Tell Halula (Arbuckle 2014; Saña Seguí 2000).

In central Anatolia, as in the Taurus of southeast Turkey, wild sheep were more numerous than goats. Sheep and a few goats were hunted at the earliest known Pre-Pottery Neolithic settlements on the Konya plain, but they formed only a modest part of a relatively diverse spectrum of species. Their presence at Boncuklu, which is at least 15 km from the nearest hill country where wild sheep were likely to have been found, suggests that small numbers of sheep may have been brought to the site and managed as a small component in a subsistence strategy that probably also involved some cultivation of cereals (Baird et al. 2018). In Cappadocia a little to the east, at the large and long-lived settlement of Aşıklı Höyük, sheep were always more important economically. In the mid-9th millennium BCE, early in the life of the settlement, the sheep were local wild animals, but, by around 8200 BCE, the age and sex profile of the sheep had changed (Stiner et al. 2014). However, the presence of numbers of foetal and neonatal remains, implying females giving birth in the settlement, and the culling of young male animals (and keeping females for breeding) are signals that some sheep were kept at the site from the earliest levels (Stiner et al. 2022). In the beginning, sheep represented about a quarter of the animals whose meat fed the population of Aşıklı Höyük. As time went by, and as methods of keeping sheep developed, sheep meat became more and more important. In the later stages of the life of the settlement the herding of sheep became a staple of the economy. The presence of their dung in areas among the houses in the earliest levels represents another line of evidence for the keeping of sheep within the settlement. Analysis of the sediments has detected concentrations of soluble sodium, chlorine, nitrate, and nitrate-nitrogen isotope attributable to their urine (Abell et al. 2019). Stiner et al. (2022) describe this earliest stage as a 'catch-and-grow' strategy: wild lambs and goat kids would be captures, brought to the settlement and raised for several months prior to slaughter. In time, people began to keep some animals and breed them; in the later levels of the site, when sheep and goat provided much the greater part of their meat, full-blown herding of large flocks of (domesticated) animals became the norm. Thus, Aşıklı Höyük, over a period of a thousand years, provides another example (like Çayönü that was mentioned above) of the local implementation of animal management strategies, leading to domestication, and ultimately to reliance on domesticated flocks (Figure 3.5).

The same cannot be said of the island of Cyprus, where the Pleistocene fauna was typically insular; it lacked many of the mainland species, but had unique versions of at least two, hippopotamus and elephant, that had evolved pygmy island populations. The earliest human presence dates to the end of the Younger Dryas phase and the very beginning of the Holocene (Simmons 1991). Traces of their presence in a collapsed rock-shelter on the south coast of the island, Aetokremnos, have proved controversial. Alan Simmons, who excavated the site, believed that a group of hunter-gatherers from the mainland had taken up residence, and were responsible for the mass slaughter of pygmy hippo, whose bones were found in large quantities. Others have wanted to separate a stratum full of hippo bones, for which there was no evidence of human slaughter or butchery, from an overlying stratum for which hearths and chipped stone artefacts documented its human occupation as a rock-shelter. This alternative reading of the stratigraphy leads to the conclusion that human hunter-gatherers were not responsible for the slaughter of the pygmy hippo, which were probably extinct before the first humans visited the island. Among the flints and ashes associated with that human occupation, however, the archaeo-zoologist Jean-Denis Vigne recognised a handful of bones of wild boar (Vigne et al. 2009). Zoologists believe that wild boar was absent from Cyprus, and Vigne therefore concluded that Palaeolithic hunter-gatherers had introduced them from the mainland. In particular, Vigne noted that these wild boars were somewhat

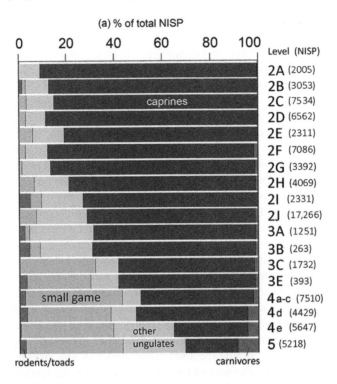

Figure 3.5 In the earliest level at Aşıklı Höyük (Level 5), there is a typical broad-spectrum range of species. Level by level at Aşıklı Höyük the caprines (mostly sheep) become more important, supplying around 90% of the meat when the settlement had its greatest population. (Stiner et al. 2014, Fig. 2, with permission)

smaller than those from contemporary sites on the nearby mainland; he therefore reasoned that a number of wild boar must have been introduced at some earlier date, allowing for the isolated population to begin to reduce in size, a characteristic of a small island population. The implication for us is that Epipalaeolithic hunter-gatherers somewhere in the north Levant or southeast Turkey already knew of Cyprus, and had deliberately taken some wild boar to the island with the intention of establishing a breeding population that would be available for hunting. It would be exciting and very helpful if a site with cultural evidence of an Epipalaeolithic group could be found somewhere on the island.

There is more to follow. As our knowledge of the early prehistory of Cyprus has developed over recent years, there is now only a small gap between the occupation of Aetokremnos around 10,000 BCE and regular Pre-Pottery Neolithic settlements in the south and southwest of the island (of which more in a later chapter): we now know of the continuous occupation of Cyprus from some time before 8500 BCE (from at least late in the early Pre-Pottery Neolithic and through the later Pre-Pottery Neolithic). Given the absence from Cyprus of the larger animals which mainland communities either hunted, managed, or were beginning to domesticate, successive groups of colonists intending to settle on the island brought breeding stock with them (Vigne et al. 2011). At the best explored and studied site of Shillourokambos, dating around 8000 BCE in the later Pre-Pottery Neolithic, dog, fox, pig, the Mesopotamian fallow deer, goat, sheep, and wild cattle were present (Vigne et al. 2000). The dates at which sheep and goat appear in Cyprus is as early the present evidence for the beginnings of domestication on the north Levantine mainland; and other species, such as cattle, pig and fallow deer, were apparently released into the wild in order to be available for hunting.

Wild cattle (Bos Taurus) were very widespread throughout Southwest Asia in the final Pleistocene. Their bones are usually rather few on most sites, but wild cattle were important contributors to the meat diet of the final Epipalaeolithic and earliest Neolithic settlements in the Euphrates valley in north Syria (Gourichon & Helmer 2008) and southeast Turkey. At early Pre-Pottery Neolithic Göbekli Tepe, for example, cattle bones were the largest component of the faunal assemblage. And it is in those regions that there are early signs of wild herd management and domestication. From long-lived settlements there is evidence of slow change of the wild cattle over time, culminating in the identification of herds of domestic cattle. Wild bulls are much larger than the cows (sexual dimorphism), and the archaeozoologists who studied the cattle bones from a cluster of sties on the Euphrates in north Syria were able to chart the slow reduction in sexual dimorphism (Helmer et al. 2005). Another study compared the cattle and red deer from Çayönü, near Diyarbakır in southeast Turkey (Hongo et al. 2009). Both species prefer a woodland environment, and both were hunted in the earliest stages of the settlement, near the end of the early Pre-Pottery Neolithic. While cattle are known to have become a domesticated form at some stage in the Pre-Pottery Neolithic, red deer were never known to be domesticated. The study compared the bones of the two species looking for changes in size, or changes in the kill pattern, and using stable isotope analyses to look for changes in diet. There was a distinct reduction in the size of cattle in the latter part of the late Pre-Pottery Neolithic, while red deer remain the same size throughout. The kill patterns of the two species also diverged over time: while red deer were killed at the same rate whatever their age, there was an increasing trend for the cattle to be killed at juvenile and sub-adult ages. Finally, the stable isotope analyses showed a gradual change in the plant diet of the cattle over time, which is interpreted as the result of the animals being moved away from their natural environment under human control. Together, these studies suggest that in that region wild cattle were being managed and herded from the early part of the later Pre-Pottery Neolithic, at very much the same time that sheep and goat were also being more intensively managed.

The hunting of wild cattle, especially wild bulls, held a special place in some communities to the extent that no attempt was made to bring them into domestication. The site of Göbekli Tepe, situated on a bare limestone ridge overlooking the great Mesopotamian plain, and full of monumental enclosures populated with tall, T-shaped monoliths, is exceptional. It dates to the early Pre-Pottery Neolithic and the beginning of the late Pre-Pottery Neolithic. The construction of the massive circular enclosures and the quarrying, moving, sculpting and erecting of the dozens of T-shaped monoliths must have required a large-scale workforce over periods of time. Bones of wild cattle dominate the faunal spectrum of Göbekli Tepe, which might be explained as the most economical way of providing food for a large workforce engaged in energy-sapping labour. However, there are other indications that together make it more likely that there was feasting at Göbekli Tepe (Dietrich et al. 2012), and the hunting of wild cattle for the feasting matches the prominence of the carved representations of wild bulls on the surfaces of the monoliths.

Another unique site with evidence of feasting on the meat of hunted wild cattle is under excavation at Kfar HaHoresh in the hills of northern Israel (Goring-Morris & Horwitz 2007), dating to the middle of the late Pre-Pottery Neolithic period. The site was not a settlement, but a place where the bodies of the dead were brought for burial. There are plaster floors and other pieces of construction, but no regular buildings. There is, however, powerful evidence of rituals in which the hunting, butchery and feasting on the meat of wild cattle played important roles. In one large pit the excavators found a succession of carefully deposited joints of wild beef representing parts of at least eight animals (Goring-Morris 2005: 906–8). All the cattle were wild aurochs (Bos primigenius).

Hunting wild cattle, feasting in large parties on the meat, and the incorporation of bull frontals with their wide horns into the interiors of houses is a prominent feature of the social life of the inhabitants of Çatalhöyük in the early seventh millennium BCE (Bogaard et al. 2009; Russell et al. 2009; Twiss et al. 2009). A study of all the recovered Bos horns has revealed a significant bias in favour of mature wild bulls (Russell et al. 2009: 29). Hunters (a group of hunters around a huge wild bull is the subject of a wall-painting) were deliberately seeking the largest and most dangerous animals, rather than culling the easier targets; and the skulls with their huge spread of horns served to commemorate the exciting rituals of hunting and feasting. Close to the end of the history of the occupation of the east mound at Çatalhöyük, as the Pre-Pottery Neolithic shaded into the late Neolithic, there was a rapid shift from away from hunting wild cattle in favour of herding domesticated beasts, a change which conforms in date with a number of other sites, as we shall see in a later chapter. With the late Neolithic and the beginnings of reliance on pottery for storage, cooking, serving, and pouring drinks, there is evidence that people had begun to make use of milk. Keeping cattle purely for their meat would hardly make sense in the Neolithic, especially where the wild aurochs was available when needed. Keeping cows for milking makes more sense; and using oxen for pulling the plough, the cart, or the sledge opens the way for more extensive and efficient farming. There is some evidence for the kind of pathological damage to the lower limbs that occurs in cattle used for traction in the final levels of Çatalhöyük East (the large Neolithic settlement), and more on the Bos bones from Çatalhöyük West (the adjacent final Neolithic into Chalcolithic settlement) (Kamjan et al. 2022). There is much better evidence for the use of milk and milk products from the lipids detected on the pottery from the final levels at Çatalhöyük and several other settlements of the late Neolithic (Evershed et al. 2008). The co-occurrence of domesticated cattle and ceramics with traces of the fatty acids associated with milk and milk products, and the association of cattle and milk products among the early

Neolithic communities across Europe, suggest the domestication of cattle was responsible for their use for both traction and for milk. Andrew Sherratt (1981) had proposed that a 'secondary products revolution' followed sometime after the primary use of domesticated flocks and herds for their meat. We can now see some of the detail of the development of these secondary products both in at the end of the later Pre-Pottery Neolithic and the late Neolithic of southwest Asia and across Europe (Marciniak 2011).

Wild pig was an even more widespread species than wild cattle, but they were hunted in some places, sometimes intensively, but scarcely at all, or not all, in other places. Similarly, domesticated pig bones have been found at some sites and are absent at others. As omnivores, pig are the odd one out among the quartet of domesticated species, which implies quite different ways of managing domesticated pig. Their appearance, whether as hunted wild boar, or as domesticated pig, at some sites and their absence at others suggests that their meat was a matter of cultural choice rather than availability, as remains the case today.

In summary, as archaeo-zoologists find more sophisticated ways to analyse the faunal assemblages, and as they have been able to work with material from a single settlement that was occupied over several or many centuries, two matters have become clear: it is a very diverse picture with no single centre from which domesticated animals were spread; and it was a long and gradual process from the beginnings of herd management to the formal recognition of domesticated status. Different regions show different patterns, even neighbouring sites can present different pictures.

Goat were intensively exploited and first became a domesticated species in the northern Zagros; local flocks of wild sheep were already being managed where the Taurus meet the Zagros in northeast Iraq at the time when the Younger Dryas period ended; sheep (and goat) were managed and became domesticated across the northern arc of the hilly flanks; wild cattle were heavily exploited in the north Levant and adjacent southeast Turkey, where they were managed and gradually emerged as a domesticated species; and a jumbled picture for wild pig and their domestication, a species that was widely available and easily domesticated, but a resource that was important in one place and apparently avoided in another in response to local cultural preferences and prohibitions. The situation of the island of Cyprus that has emerged over the last two decades is more indicative than many have realised (or perhaps I am biased, because it was where I first became involved with the problem of our ignorance of the Neolithic). Ben Arbuckle (2014) summarised well the recent state of our knowledge, which has continued to move forward, as some of the references in previous paragraphs have illustrated. It is now clear that the beginnings of animal management date much earlier than archaeologists had imagined was possible. And the mosaic picture that is emerging shows that the idea that animal domestication began in one place and spread around the rest of the hilly flanks zone was mistaken.

But domestication is not the end of the story: people did not stop hunting because they had begun to herd sheep and goat. A group of archaeo-zoologists who had assembled the evidence for the beginnings of animal husbandry at sites in the northern Levant noted that domesticated sheep and goat accounted for less than 30% of all the animal bones at settlements in the centuries following domestication (Peters et al. 1999). A group of researchers who had assembled a database of all the faunal data recovered from 114 archaeological sites across Southwest Asia and Southeast Europe dating between 10,000 BCE (the end of the Epipalaeolithic and the beginning of the Southwest Asia Neolithic) and 5500 BC. They separated out five regional groups within the data from Southwest Asia: Central Anatolia, Cyprus, the upper Euphrates region in north Syria and southeast Turkey, a Tigris-Zagros region, and the southern Levant. Their analyses demonstrated the considerable variations in patterns of exploitation, as discussed in some detail by Arbuckle (2014), and some surprising facts emerged. At settlements in the Euphrates region, for example, there were domesticated sheep, goat, cattle and pig by or soon after 8500 BCE, but their bones accounted for less than 10% of the total assemblage size. By the middle of the late Pre-Pottery Neolithic the percentage of domesticated animal bones had risen to about 40%, and to about 45% by the end of the Pre-Pottery Neolithic. A similar picture emerged for their Tigris-Zagros region. The southern Levantine sites, however, produced a very different picture: the contribution of domestic animals remained as low as 1% of the total animal bones until quite late in the late Pre-Pottery Neolithic, when the proportion increased dramatically to around 35%. In short, dependence on herds of domesticated animals emerged slowly, and it was only in the later, pottery Neolithic that communities obtained more than half of their meat from their domesticated flocks and herds.

References

Abell, J. T., Quade, J., Duru, G., Mentzer, S. M., Stiner, M. C., Uzdurum, M., & Özbaşaran, M. (2019). Urine Salts Elucidate Early Neolithic Animal Management at Aşıklı Höyük, Turkey. *Science Advances*, 5(4), eaaw0038. doi:10.1126/sciadv.aaw0038

Arbuckle, B. S. (2014). Pace and Process in the Emergence of Animal Husbandry in Neolithic Southwest Asia. *Bioarchaeology of the Near East*, 8, 53–81.

Baird, D., Fairbairn, A., Jenkins, E., Martin, L., Middleton, C., Pearson, J., . . . Elliott, S. (2018). Agricultural Origins on the Anatolian Plateau. *Proceedings of the National Academy of Sciences*, 115(14), E3077–86. doi:10.1073/pnas.1800163115

Binford, L. R. (1968). Post-Pleistocene Adaptations. In L. R. Binford & S. R. Binford (Eds.), *New Perspectives in Archaeology* (pp. 313–41). Chicago, IL: University of Chicago Press.

Bogaard, A., Charles, M., Twiss, K. C., Fairbairn, A., Yalman, N., Filipovic, D., . . . Henecke, J. (2009). Private Pantries and Celebrated Surplus: Storing and Sharing Food at Neolithic Catalhoyuk, Central Anatolia. *Antiquity*, 83(321), 649–668.

Daly, K. G., Maisano Delser, P., Mullin, V. E., Scheu, A., Mattiangeli, V., Teasdale, M. D., . . . Bradley, D. G. (2018). Ancient Goat Genomes Reveal Mosaic Domestication in the Fertile Crescent. *Science*, 361(6397), 85–8. doi:10.1126/science.aas9411

Davis, S. J. M. (1983). The Age Profiles of Gazelles Predated by Ancient Man in Israel: Possible Evidence for a Shift from Seasonality to Sedentism in the Natufian. *Paléorient*, 9, 55–62.

Davis, S. J. M. (2005). Why Domesticate Food Animals? Some Zoo-Archaeological Evidence from the Levant. *Journal of Archaeological Science*, 32(9), 1408–16. doi:10.1016/j.jas.2005.03.018|ISSN 0305-4403

Davis, S. J. M., Lernau, O., & Pichon, J. (1994). The Animal Remains: New Light on the Origin of Animal Husbandry. In M. Lechevallier & A. Rosen (Eds.), *Le gisement de Hatoula en Judée occidentale, Israël* (pp. 83–100). Paris: Mémoires et Travaux du Centre de Recherche Français de Jerusalem, No. 8. Association Paléorient.

Davis, S. J. M., & Valla, F. R. (1978). Evidence for Domestication of Dog 12,000 Years Ago in Natufian of Israel. *Nature*, 276(5688), 608–10.

Dietrich, O., Heun, M., Notroff, J., Schmidt, K., & Zarnkow, M. (2012). The Role of Cult and Feasting in the Emergence of Neolithic Communities. New evidence from Göbekli Tepe, South-Eastern Turkey. *Antiquity*, 86(333), 674–95.

Ducos, P. (1993). Some remarks about Ovis, Capra, and Gazelle remains from two PPNB sites from Damascene, Syria, Tell Aswad and Ghoraife. In H. Buitenhuis & A. T. Clason (Eds.), *Archaeozoology of the Near East* (pp. 37–42). Leiden: Universal Book Services.

Evershed, R. P., Payne, S., Sherratt, A. G., Copley, M. S., Coolidge, J., Urem-Kotsu, D., . . . Burton, M. M. (2008). Earliest Date for Milk Use in the Near East and Southeastern Europe Linked to Cattle Herding. *Nature*, 455(7212), 528–31. doi:10.1038/nature07180

Flannery, K. V. (1969). The Origins and Ecological Effects of Early Domestication in Iran and the Near East. In P. J. Ucko & G. W. Dimbleby (Eds.), *The Domestication and Exploitation of Plants and Animals* (pp. 73–100). London: Duckworth.

Goring-Morris, A. N. (2005). Life, Death and the Emergence of Differential Status in the Near Eastern Neolithic: Evidence from Kfar HaHoresh, Lower Galilee, Israel. In J. Clarke (Ed.), *Archaeological Perspectives on the Transmission and Transformation of Culture in the Eastern Mediterranean* (pp. 89–105). Oxford: Council for British Research in the Levant & Oxbow Books.

Goring-Morris, A. N., & Horwitz, L. K. (2007). Funerals and Feasts during the Pre-Pottery Neolithic B of the Near East. *Antiquity*, 81(314), 902–19. Retrieved from http://antiquity. ac.uk/ant/081/ant0810902.htm

Gourichon, L., & Helmer, D. (2008). Étude archéozoologique de Mureybet. In J. J. Ibáñez (Ed.), *Le site Néolithique de Tell Mureybet (Syrie du Nord). En hommage à Jacques Cauvin* (pp. 115–228). Oxford: British Archaeological Reports, International Series 1843, Archaeopress.

Helmer, D., Gourichon, L., Monchot, H., Peters, J., & Saña Seguit, M. (2005). Identifying Early Domestic Cattle from Pre-Pottery Neolithic Sites on the Middle Euphrates Using Sexual Dimorphism. In J. D. Vigne, D. Helmer, & J. Peters (Eds.), *9th International Council of Archaeozoology Conference*, Durham 2002 (pp. 86–95). Oxford: Oxbow Books.

Helmer, D., Gourichon, L., & Stordeur, D. (2004). À l'aube de la domestication animale. Imaginaire et symbolisme animal dans les premières sociétés néolithiques du nord du Proche-Orient. *Anthropozoologica*, 39(1), 143–63.

Hole, F., & Flannery, K. (1968). The Prehistory of Southwestern Iran: A Preliminary Report. *Proceedings of the Prehistoric Society*, 33, 147–206.

Hole, F., Flannery, K. V., & Neely, J. A. (1969). *Prehistory and Human Ecology of the Deh Luran Plain; an Early Village Sequence from Khuzistan, Iran. Memoirs of the Museum of Anthropology, University of Michigan, no. 1*. Ann Arbor: Museum of Anthropology, University of Michigan.

Hongo, H., Meadow, R. H., Öksüz, B., & Gülçin, I. (2004). Animal Exploitation at Çayönü Tepesi, Southeastern Anatolia. *TÜBA-AR*, 7, 107–19.

Hongo, H., Meadow, R. H., Ōksuz, B., & Ilgesdi, G. (2002). The Process of Ungulate Domestication in Prepottery Neolithic Çayönü, Southeastern Turkey. In H. Buitenhuis, A.

M. Choyke, M. Mashkour, & A. H. Al-Shiyab (Eds.), *Archaeozoology of the Near East IV* (pp. 153–65). Groningen: ARC Publications No. 62.

Hongo, H., Meadow, R. H., Öksüz, B., & Ilgezdi, G. L. I. (2005). Sheep and Goat Remains from Çayönü Tepesi, Southeastern Anatolia. In H. Buitenhuis, A. Choyke, L. Martin, L. Bartosiewicz, & M. Mashkour (Eds.), *Archaeozoology of the Near East VI* (pp. 112–23). Groningen: ARC Publication 123.

Hongo, H., Pearson, J., Öksüz, B., & Ilgezdi, G. (2009). The Process of Ungulate Domestication at Çayönü, Southeastern Turkey: A Multidisciplinary Approach Focusing on Bos sp. and Cervus elaphus. *Anthropozoologica*, 44(1), 63–78.

Kamjan, S., Erdil, P., Hummel, E., Çilingiroğlu, Ç., & Çakırlar, C. (2022). Traction in Neolithic Çatalhöyük? Palaeopathological Analysis of Cattle and Aurochs Remains from the East and West Mounds. *Journal of Anthropological Archaeology*, 66, 101412. doi:10.1016/j.jaa.2022.101412

Legge, A. J., & Rowley-Conwy, P. A. (2000). The Exploitation of Animals. In A. M. T. Moore, G. C. Hillman, & A. J. Legge (Eds.), *Village on the Euphrates*. Oxford: Oxford University Press.

Marciniak, A. (2011). The Secondary Products Revolution: Empirical Evidence and Its Current Zooarchaeological Critique. *Journal of World Prehistory*, 24, 117–30. doi:10.1007/s10963-011-9045-7

Perkins, D. (1964). Prehistoric Fauna from Shanidar, Iraq. *Science*, 144, 1565–6.

Peters, J., Helmer, D., Von Den Driesch, A., & Sana Segui, A. (1999). Early Animal Husbandry in the Northern Levant. *Paléorient*, 25, 27–47.

Rosenberg, M. N. (2011). Hallan Çemi. In M. Özdoğan, N. Başgelen, & P. Kuniholm (Eds.), *The Neolithic in Turkey. New Excavations and New Research: The Tigris Basin* (pp. 61–78). İstanbul: Arkeoloji ve Sanat Yayinlari.

Russell, N., Martin, L., & Twiss, K. C. (2009). Building Memories: Commemorative Deposits at Çatalhöyük. *Anthropozoologica*, 44(1), 103–25. doi:10.5252/az2009n1a5

Saña Seguí, M. (2000). Animal Resource Management and the Process of Animal Domestication at Tell Halula (Euphrates Valley-Syria) from 8800 BP to 7800 BP. In M. Mashkour, A. M. Choyke, H. Buitenhuis, & F. Poplin (Eds.), Archaeozoology of the Near East IVA (pp. 241–56). Groningen: ARC Publications.

Sherratt, A. (1981). Plough and Pastoralism: Aspects of the Secondary Products Revolution. In I. Hodder, G. L. Isaac, & N. Hammond (Eds.), *Pattern of the Past* (pp. 261–306). Cambridge: Cambridge University Press.

Simmons, A. (1991). Humans, Island Colonization and Pleistocene Extinctions in the Mediterranean: The View from Akrotiri Aetokremnos, Cyprus. *Antiquity*, 65, 857–69.

Smith, A., Oechsner, A., Rowley-Conwy, P., & Moore, A. M. T. (2022). Epipalaeolithic Animal Tending to Neolithic Herding at Abu Hureyra, Syria (12,800–7800 calBP): Deciphering Dung Spherulites. *PLoS ONE*, 17(9), e0272947. doi:10.1371/journal.pone.0272947

Starkovich, B., & Stiner, M. (2009). Hallan Çemi Tepesi: High-Ranked Game Exploitation alongside Intensive Seed Processing at the Epipaleolithic-Neolithic Transition in Southeastern Turkey. *Anthropozoologica*, 44(1), 41–61.

Stiner, M. C., Buitenhuis, H., Duru, G., Kuhn, S. L., Mentzer, S. M., Munro, N. D., . . . Özbaşaran, M. (2014). A Forager–Herder Trade-Off, from Broad-Spectrum Hunting to Sheep Management at Aşıklı Höyük, Turkey. *Proceedings of the National Academy of Sciences of the United States of America*, 111(23), 8404–9. doi:10.1073/pnas.1322723111

Stiner, M., & Munro, N. D. (2002). Approaches to Prehistoric Diet Breadth, Demography, and Prey Ranking Systems in Time and Space. *Journal of Archaeological Method and Theory*, 9(2), 175–208.

Stiner, M. C., Munro, N. D., Buitenhuis, H., Duru, G., & Özbaşaran, M. (2022). An Endemic Pathway to Sheep and Goat Domestication at Aşıklı Höyük (Central Anatolia, Turkey). *Proceedings of the National Academy of Sciences*, 119(4), e2110930119. doi:10.1073/pnas.2110930119

Stutz, A. J., Munro, N. D., & Bar-Oz, G. (2009). Increasing the Resolution of the Broad Spectrum Revolution in the Southern Levantine Epipaleolithic (19–12 ka). *Journal of Human Evolution*, 56(3), 294–306.

Tchernov, E., & Valla, F. (1997). Two New Dogs, and Other Natufian Dogs, from the Southern Levant. *Journal of Archaeological Science*, 24(1), 65–95.

Twiss, K. C., Bogaard, A., Charles, M., Henecke, J., Russell, N., Martin, L., & Jones, G. (2009). Plants and Animals Together: Interpreting Organic Remains from Building 52 at Çatalhöyük. *Current Anthropology*, 50(6), 885–95. doi:10.1086/644767

Vigne, J.-D., Carrère, I., Briois, F., & Guilaine, J. (2011). The Early Process of Mammal Domestication in the Near East: New Evidence from the Pre-Neolithic and Pre-Pottery Neolithic in Cyprus. *Current Anthropology*, 52(S4), S255–71. Retrieved from http://www.jstor.org/stable/10.1086/659306

Vigne, J.-D., Carrère, I., Saliège, J.-F., Person, A., Bocherens, H., Guilaine, J., & Briois, F. (2000). Predomestic Cattle, Sheep, Goat and Pig during the Late 9th and the 8th Milleniun cal. BC on Cyprus: Preliminary Results of Shillourokambos (Parekklisha, Limassol). In M. Mashkour, A. M. Choyke, H. Buitenhuis, & F. Poplin (Eds.), *Archaeozoology of the Near East IV. Proc. 4th International Symposium on the Archaeozoology of Southwestern Asia and Adjacent Areas* (Vol. ARC publicaties, pp. 52–75). Groningen: Centre for Archaeological Research and Consultancy.

Vigne, J.-D., Zazzoa, A., Saliege, J.-F., Poplin, F., Guilaine, J., & Simmons, A. (2009). Pre-Neolithic Wild Boar Management and Introduction to Cyprus More Than 11,400 Years Ago. *Proceedings of the National Academy of Sciences of the United States of America*, 106(38), 16135–8.

Watkins, T., Dobney, K., & Nesbitt, R. M. (1995). *Qermez Dere, Tel Afar; Interim Report No. 3*. Edinburgh: Department of Archaeology, University of Edinburgh.

Yeomans, L., Martin, L., & Richter, T. (2019). Close Companions: Early Evidence for Dogs in Northeast Jordan and the Potential Impact of New Hunting Methods. *Journal of Anthropological Archaeology*, 53, 161–73. doi:10.1016/j.jaa.2018.12.005

Zeder, M. A. (2011). The Origins of Agriculture in the Near East. *Current Anthropology*, 52(S4), S221–35. Retrieved from http://www.jstor.org/stable/10.1086/659307

Zeder, M. A. (2012). The Broad Spectrum Revolution at 40: Resource Diversity, Intensification, and an Alternative to Optimal Foraging Explanations. *Journal of Anthropological Archaeology*, 31(3), 241–64. doi:10.1016/j.jaa.2012.03.003

Zeuner, F. E. (1955). The Goats of Early Jericho. *Palestine Exploration Quarterly*, 87(1), 70–86. doi:10.1179/peq.1955.87.1.70

4   Early Epipalaeolithic – The Transformation Begins

This chapter is the first of a sequence of five that dive into the archaeology of the ENT, the Epipalaeolithic–Neolithic transformation. We will see how the social, cultural and economic lives of the people who lived around the hilly flanks of the Fertile Crescent changed very gradually at first, but with increasing tempo and significance. Where does the Epipalaeolithic of Southwest Asia come from, and why is it important for us? In the third part of the book I will be setting the Epipalaeolithic–Neolithic transformation in Southwest Asia into its place in the 'deep history' of the evolution of human culture and society. It has been common to find the late Epipalaeolithic, specifically the Natufian of the southern Levant, used as a sort of prelude against which to set the drama of the Neolithic. In this account of the transformation, I want us to see the whole of the Epipalaeolithic as the first act of a four-act drama (the Neolithic consisting of three acts).

First, at the start I should justify why this ENT begins with the transition from the Upper Palaeolithic to the Epipalaeolithic. We should first consider, therefore, on what basis these Epipalaeolithic developments were novel. When the Upper Palaeolithic begins in Southwest Asia, around 50,000 years ago, the ambivalent situation of the later Middle Palaeolithic period has been resolved. From around 120,000 years ago at sites in the Levant the puzzling fossil remains of two species have long been known (but have only recently been dated): the skeletal remains of Neanderthals and Homo sapiens are represented, but accompanied by a single material culture (that is, a common chipped stone tool-making tradition) and a single cultural tradition of burying the dead in caves and rock-shelters associated with everyday occupation (Tillier & Arensburg 2017). It used to be thought that the Homo sapiens finds from the Mt Carmel caves represented a first but ephemeral expansion of the species beyond Africa; closer to the end of the Middle Palaeolithic only Neanderthal skeletal remains were known – until recently. Now, thanks to the discovery of new Homo sapiens remains in Israel dating to 60–50,000 years ago (Israel Hershkovitz et al. 2015), we can be reasonably sure that the two groups co-existed and, given their shared material culture, that they communicated with each other. Indeed, the recent reconstruction of the Neanderthal genome has enabled scientists to hypothesise a model of multiple episodes of Neanderthal gene flow into both European and Asian sapiens populations (Kuhlwilm et al. 2016; Villanea & Schraiber 2019).

DOI: 10.4324/9781351069281-5

Whatever the complexities of the later Middle Palaeolithic, from the beginning of the Upper Palaeolithic period a single population of Homo sapiens is present, with a new chipped stone tool-making industry. The Upper Palaeolithic presents us with a scenario in the Levant in which for the first time a second chipped stone industry appears alongside the first; we seem to have evidence that, for some time, there were two contemporary populations living as mobile forager bands within the same environment with somewhat different cultural traditions of stone tool-making. The Upper Palaeolithic millennia in the Levant are characterised by small, highly mobile forager bands living at low density (Anna Belfer-Cohen & Goring-Morris 2017). Taking the Upper Palaeolithic in general as the baseline, a simple index that distinguishes the Epipalaeolithic from the Upper Palaeolithic period was defined a long while ago; it was the trend towards miniaturisation in the chipped stone industry. At its simplest, the long, narrow, parallel-sided blades of the Upper Palaeolithic period became bladelets, from which large numbers of microliths (worked segments that have been snapped from bladelets and retouched into shape) were produced. Radiocarbon dates have shown that the changes in chipped stone manufacture that conventionally distinguish the Epipalaeolithic from the Upper Palaeolithic occur around 23,000 years ago.

Thus, the Epipalaeolithic now spans from the Last Glacial Maximum through the Bølling-Allerød climatic recovery to the end of the Younger Dryas phase almost 12,000 years ago. Prehistoric archaeologists once thought that the Upper Palaeolithic was a long period, and the Epipalaeolithic was a mere transitional epilogue at the end of the Palaeolithic, but the radiocarbon dates have shown the true extent of the later period. We can now see that the Upper Palaeolithic lasted a little more than 25,000 years, within which timespan archaeologists can detect very little change or development in tool-making techniques or preferences. By contrast, across the Epipalaeolithic of approximately eleven millennia, there was an appreciable pace of cultural change in chipped stone tool-making. More significantly, Epipalaeolithic communities changed their settlement strategies, living together in larger numbers, becoming less mobile and more sedentary, changing their subsistence modes to engage in the annual harvesting and storage of grass seeds, cereals and pulses, complemented that with a strategy of broad-spectrum hunting that included many smaller species.

Now that it is clear that there were two industrial traditions operating in the Upper Palaeolithic, we can say that the Epipalaeolithic small-scale chipped stone tradition developed out of the Ahmarian of the Upper Palaeolithic (Belfer-Cohen & Goring-Morris 2018; Goring-Morris & Belfer-Cohen 2017). People continued to prepare their cores in the same way, but the cores and therefore the blades struck from them were smaller. The convention today is that when a blade is less than 40 mm long and/or less than 12 mm wide it is a bladelet, and the Epipalaeolithic industries are essentially bladelet industries in which only a minority of pieces are blades or made from blades. Microliths are formed from bladelets and come in many shapes but always in a very small size, with miniaturised retouch in scale with that size. Archaeologists often refer to triangular and trapeze-shaped microliths as 'geometric', and the rest as 'non-geometric'.

In most parts of southwest Asia our knowledge of this period is still very sketchy, simply because too few sites have been located and investigated. Only in Israel, Jordan and, to a lesser extent, Syria have sufficient sites been investigated to allow the reconstruction of a cultural sequence, so we should start there. At the end of the chapter we shall expand our perspective to look at other areas of southwest Asia where there are Epipalaeolithic sites. In the Levantine corridor, where we have relatively detailed knowledge and radiocarbon dates, we can divide the Epipalaeolithic into two sub-periods, an early Epipalaeolithic (that includes what the specialists call the Early and the Middle Epipalaeolithic), which is the subject of this chapter, and a late Epipalaeolithic, a much shorter period that requires a chapter of its own.

The Epipalaeolithic Sequence in the Levantine Corridor

As with the earlier periods, the cultural sequence was based on comparative stratigraphy initially constructed by Dorothy Garrod (1937) and René Neuville (1934) on the basis of work in what was then the British mandate of Palestine. As with the Upper Palaeolithic, we now know that the Epipalaeolithic sequence is not as they saw it, the simple, unilinear progression of a succession of chipped stone industries. Extensive research work in Jordan east of the Jordan valley, and, since 1967, in Sinai and the Negev, has changed that simple succession. Taken together with the radiocarbon dating evidence, we now know that there were contemporary localised and minimally differentiated cultural groups in neighbouring areas, and in some cases different cultural groups apparently occupying the same region at the same time (Figure 4.1).

After a dangerously expansive phase of identifying and naming all sorts of new 'cultures' or 'industries' in the late 1960s and 1970s (when there was a rapid expansion in the number of people specialising in the field in the southern Levant), the tendency has been steadied, and for some time there has been agreement among those active in Epipalaeolithic studies in the Levant to operate in terms of five or six groups defined almost exclusively in terms of their chipped stone traditions. The additional complexity arises from the recognition that, within the southern Levant, there were different ecological zones within which Epipalaeolithic hunter-gatherer groups necessarily pursued specialised subsistence strategies, which were matched by refined specialisation in their chipped stone tool-kits (Byrd et al. 2016). Three of them are the 'old' cultures identified and stratified in the Carmel caves, and they form the core of the chronological sequence. They are the Kebaran, which dates to the Last Glacial Maximum (LGM) climatic phase, the Geometric Kebaran, which dates to the phase of Bølling-Allerød climatic amelioration following the LGM, and the Natufian, which dates to the last three millennia of the Epipalaeolithic period, which includes the millennium-long Younger Dryas phase. For the purposes of this chapter, and for the sake of simplicity, we can continue to use the old terms.

Byrd et al. (2016) concentrate on the early Epipalaeolithic and particularly the first half of that period, during the Last Glacial Maximum. The easiest ecological

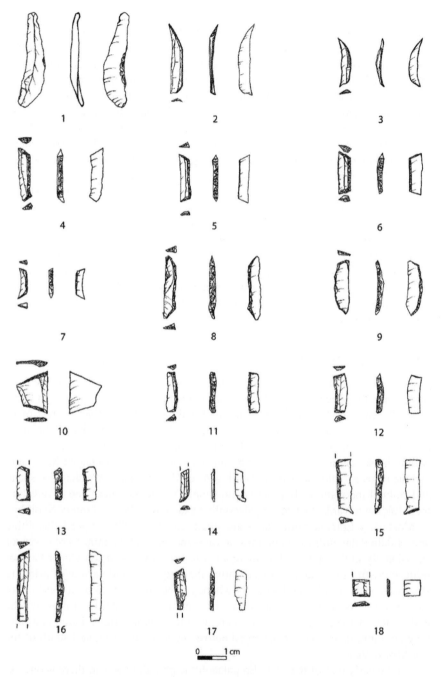

Figure 4.1 Microlithic tools from Neve David. Note the 1 cm scale. (Liu et al. 2020, Fig. 4, with permission)

distinction to recognise within the southern Levant is between the Mediterranean woodland zone on either side of the Jordan valley, and the semi-arid steppe and arid regions of southern and eastern Jordan, the Negev, and Sinai. The semi-arid and arid parts of the region would have offered less in the way of plant resources, and groups living in those environments would probably have relied more on greater mobility and hunting than groups living in the Mediterranean woodland zone. The reconstructed daily foraging range of groups in the east and south would also have necessarily been larger (Byrd et al. 2016). On the other hand, the Mediterranean woodland zone was relatively homogeneous, which would have required forager bands to move relatively frequently. The situation in the rift valley itself was different again; the valley bottom, the adjacent slopes, and then the steep sides to the valley offered varied ecological zones in close proximity to one another, which would allow a hunter-gatherer group to access different, complementary resources from a single base-camp. It follows that the tool-kits of contemporary groups living in somewhat different ways in the different environments might vary, and that the residues that groups have left for archaeologists to excavate might also vary relative to the length of occupation and the quantities and variety of activities pursued on any occupation site. In short, early Epipalaeolithic hunter-gatherer groups adapted very neatly to the different environmental zones and the resources that they offered.

The Early Epipalaeolithic Mosaic

There is variation in space and time within the early Epipalaeolithic, as we have seen. While the specialists can differentiate contemporary cultural groups in terms of small details of their chipped stone tool-making traditions that the rest of us would not be able to notice, we can afford to overlook the various labels that have been devised for these industries. For us, there are three environmental zones within the southern Levant, and two climatic stages, the Last Glacial Maximum and the recovery into the warmer Bølling-Allerød interstadial. We shall see social and cultural innovations, but there is no simple, single, unilinear line of development leading towards the late Epipalaeolithic and on into the Pre-Pottery Neolithic.

Most early Epipalaeolithic sites are small, contained, like Upper Palaeolithic and Middle Palaeolithic occupations, to rock-shelters and the mouths of caves, or occasionally in the open. While there are a few grinding slabs on Upper Palaeolithic sites, from the early Epipalaeolithic the frequency and variety of grinding stones, mortars, pestles and stone bowls begins to increase (Wright 1991). Not surprisingly, given the more generous environmental conditions following the end of the Last Glacial Maximum, there are more Geometric Kebaran sites and they are more extensively distributed across the semi-arid east and south of the southern Levant.

For the early part of the early Epipalaeolithic period, however, there is one exceptional site, Ohalo II, at the southern edge of the Sea of Galilee (Lake Kinneret) that demands attention (Nadel 2017; Nadel & Werker 1999). The site is firmly dated by radiocarbon at around 23,000 years ago, that is fully in the middle of

Figure 4.2 Part of the site of Ohalo II. Two roughly circular huts are close together; the occupation deposits on their hollowed floors are still to be excavated and they show as light grey. The light wooden wall shows as a darker outline, because, at the end of its life, each hut was burnt. The almost black patch is a large communal hearth. The stony strip at the top right corner is part of a linear dumping of refuse away from the living area. (Excavation photograph, with thanks to Dani Nadel)

the Last Glacial Maximum. It was established on the marls of the retreating Lake Lisan, the name given to the inland sea that once filled much of the rift valley that is now the Sea of Galilee, the Jordan valley and the Dead Sea. At that time the Sea of Galilee was a little shallower, and there was a short distance between its southern outflow and the north end of Lake Lisan. Ohalo II today is normally submerged under more than 2 metres of water, and was discovered in 1989 when the Sea of Galilee was reduced to an extraordinarily low level by drought compounded by the extraction of water for irrigation. Naturally, excavation opportunities have been limited, but there was another drought that exposed the site between 1999 and 2001 (Figure 4.2).

The site is exceptional for at least two reasons. The site has been submerged under water and a layer of silty lake sediment, which has ensured the unique survival of quantities of organic materials, including large amounts of plant-food materials. Secondly, as Byrd et al. (2016) have pointed out, the environmental conditions in the rift valley were especially kind during the Last Glacial Maximum. The group of fisher-hunter-gatherers who occupied Ohalo II also capitalised on its location, from which there was immediate or close access to a range of ecological zones – the waters of the Sea of Galilee, the water's edge, the low-lying lands of the rift valley, and the nearby hills to east and west – with their complementary resources.

In many ways, Ohalo II is a precocious precursor of the developed later Epipalaeo-lithic period.

The area of the site that has been investigated consists of a cluster of huts with scooped out floors, and superstructures built of wood and brush; there are several external hearths and fire-pits, an area where domestic rubbish was deposited, and a single grave at a little distance from the huts. The huts were slightly oval in plan, between three and four metres across. All of the huts had been deliberately burned, the collapsed remains sealing all the material that was on the floor when the hut was abandoned. One of the huts has been particularly closely examined (Nadel et al. 2004; Nadel & Werker 1999; Weiss 2009; Weiss et al. 2008). Its wall was formed of quite thick branches of tamarisk and willow that were set about 200 mm into the ground. Smaller branches of oak were used, together with saltbush and *Pros-opis* bushes. Remains of straw were also identified on the slightly scooped floor, which had been remade twice. The excavators have reported that they encountered a wealth of finds on the floor, including chipped stone, animal bone and preserved seeds and fruits of plants, enabling them to place within the hut the location of various domestic and house-keeping activities; although the interior was a single space, it was used in a very orderly way, and over a period of time that required its floor being renewed (Figure 4.3).

Figure 4.3 Reconstruction drawing of a brush hut at Ohalo II. The excavators found that, because the interiors were small in area, the domestic activities that took place were clearly differentiated. (Nadel & Werner 1999, Fig. 8, with thanks to Dani Nadel)

The single burial found so far is that of a male of 35–40 years of age, 1.73 m tall. He was buried lying almost flat, his head supported by three stones (Hershkovitz et al. 1995). His legs were folded back tightly at the knees so that the heels were in contact with the buttocks, and his forearms lay across his chest. Close to the head the excavators found a small implement made from a gazelle's limb bone that had been decorated with many close incisions. There are signs that this person suffered some physical disabilities in the latter years of his life to the extent that he would not have been a contributing member of his social group.

More than 150,000 seeds and fruits have been recovered from the site, revealing that 142 plant taxa were being collected. The people who lived at Ohalo II gathered wild barley and emmer, many species of small-seeded wild grasses, acorns, almonds, Pistacia, wild olives and figs, together with a range of legumes and other plants. From the surface of a grinding slab on the floor of one of the huts, traces of starch were recovered, showing for the first time (and at an extraordinarily early date) that the heavy stone grinding implements were indeed used in producing meal and flour (Piperno et al. 2004). The diversity of plants shows that people were collecting food plants across the full range of altitude, from the shores of Lake Lisan up to more than 1000 m above sea level. Gazelle were hunted in numbers, and fish from the lake were also important. Other mammals that were exploited include fallow deer, fox and hare, and a wide variety of birds. The seasons of occupation indicated by the plant remains extend through the spring, summer and autumn, and the evidence of the cementum growth in the gazelle teeth, together with the indications from the bird bones, some of which were migrant species, suggest year-round occupation. We cannot know, of course, whether the group lived there continuously throughout any calendar year; they may have used the site repeatedly at different seasons of the year over a period of years or decades, but the implication of their use of annual harvests of small-seeded grasses and cereals is that the group was in residence for months at a time.

Although the people of Ohalo II were harvesting the seeds of as many as 150 taxa, they were predominantly interested in small-seeded grasses and the cereals barley and wheat. There are far fewer grains of barley and wheat than of small-seeded grasses, but the much greater size of the minority of cereal grains increases their nutritional value. The intensive study of the grains, fragments of the cereals, and other seeds classed as weeds of cultivation, has even led the Ohalo team to claim that these wild plants were beginning to be tended and cultivated (Nadel et al. 2012; Snir et al. 2015a, 2015b).

Here, in the rift valley, at the beginning of the Epipalaeolithic period, we have a type-site for Flannery's broad-spectrum revolution (Weiss et al. 2004), the exploitation of the full range of animals, including small animals, birds, tortoises, and lots of fish, the extensive harvesting of storable plant foods, and the semisedentary occupation of a settlement located at a place from which a variety of complementary ecological zones could be exploited; all these are features that we shall find are characteristic of the Epipalaeolithic period, particularly the late Epipalaeolithic, and they are also features that are rare or absent in the Upper Palaeolithic period.

Most Geometric Kebaran sites in the Mediterranean woodland zone are small and are the result of periodic, brief re-occupations, but in the archaeologically fertile area around Mount Carmel in the north of Israel, there is a significant site on the outskirts of modern Haifa that illustrates well this tendency to larger aggregations and longer occupations at base-camp sites. Neve David is situated at the foot of the western scarp of Mount Carmel, at the mouth of a valley with a seasonal stream (Kaufman 1989). It is only a kilometre from the present coastline, but in final Pleistocene times the site looked out onto a coastal plain that was 10–12 km wide. This is the classic ecotone situation favoured by hunter-gatherers seeking to reduce their mobility: sitting on the boundary between contrasting ecological zones, the limestone massif of Mount Carmel and its seasonally dry valleys on the one hand and the Mediterranean coastal plain on the other, the occupants had ready access to the complementary resources of different ecological zones.

The estimated extent of the occupation is about 1000 m^2, which is significantly larger than most sites of the period, whose average is more like 400 m^2. More than a metre of archaeological deposit was accumulated on this open site, which implies not only repeated use over a long period but also a kind of use that brought soil and clay to the site and that prevented its easy erosion by wind and weather between occupations. Although the remains of structures may no longer be traceable, it is hard to see how the archaeological deposit can have accumulated without some use of mud as a building material. Base-camp occupation is also indicated by the density and variety of cultural and other debris. Kaufman (1989: 277) noted in particular the quantity and variety of ground stone implements, some in local limestone, the rest of black basalt obtained from sources that are tens of kilometres distant. The minute analysis of the chipped stone recovered in closely documented work at the site in 2014 provided evidence of increasing intensity of occupation through time (Liu et al. 2020).

Within the quite small area of the excavation (about 40 m^2), two burials were encountered. One of them was poorly preserved because of damage from pits dug from the much later occupation (late fourth or third millennium BC) of the site (Kaufman & Ronen 1987). The other burial was that of the tightly contracted body of a male between 23 and 30 years of age. The grave was constructed of two large stone slabs at the head and foot and smaller stones along the longer sides, forming a cist about 1.2 by 0.5 m. Over the head of the dead man a large stone mortar had been inverted. At the back of his neck part of a broken basalt bowl had been placed, and there was a small piece of a flat basalt grinding slab between the thighs. In the context of the fragmentary bowl and grinding stone it is of interest to note that the large mortar had been used until the bottom of the stone had been broken through; it too was, if not broken, then certainly exhausted. In the way that the grave was carefully constructed and the burial accompanied by (broken or exhausted) ground stone implements, this burial at Neve David (and others at other sites – see below) presages the care and rituals that often accompany the burials of the late Epipalaeolithic.

There are extraordinary sites – in the sense that they were completely unexpected and contradictory to what was thought to be the norm – in the semi-arid of north and northeast Jordan, associated with the Azraq basin in the heart of the basalt landscapes of the 'black desert'. This is an area with springs and wetlands. Two of the sites in the Azraq basin have been characterised as aggregation sites, that is locations where large numbers of mobile hunter-gatherer bands repeatedly came together in very large gatherings. The site of Kharaneh IV dates to the early Epipalaeolithic; it was first occupied in the early, Kebaran phase, but its prime use was in the Geometric Kebaran. It was first excavated by the Jordanian archaeologist Muhajed Muheisen (1988), but since 2007 a team led by Lisa Maher has undertaken detailed investigations (Macdonald et al. 2018; Macdonald & Maher 2022; Maher et al. 2012, 2016). The site extends to more than 21,000 square metres, and has a stratified occupation deposit up to 2 m thick. A programme of radiocarbon dating has shown that the site was in use over a period of around 1200 years, between around 17,800 and 16,600 BCE. The chipped stone assemblages show that the site was first in use by groups characterised as Kebaran and it continued to be used by Geometric Kebaran groups. The thick occupation deposit is rich in assemblages of stone tools, worked bone objects, red ochre, and marine shell beads that indicate a continuous sequence of repeated seasonal occupations. The stratigraphy is generally made up of thin, compacted occupation surfaces and thicker midden deposits full of chipped stone and animal bone. The density and variety of cultural material indicate that people were carrying out all of the normal activities of life, and spending substantial periods of time in residence: the aggregation sites brought together very substantial numbers of people, and they were not gathering for something like a weekend break.

There are remains of built structures, of which two have been carefully excavated (Maher et al. 2012). It is worth spending some time with these structures, because, however flimsy they may have been in structural terms, they clearly meant a good deal more to their occupants than simply shelters from the elements. Engagement and identification with their houses and settlements is a characteristic that has been discussed over many years for the early Neolithic (e.g. Hodder 1990; Watkins 1990), but Kharaneh IV shows us a much earlier example of such deep attachment to place. The structures were formed in shallow oval depressions dug into earlier archaeological deposits. The structures themselves were formed of wood and brush-wood, which, in one case, was found as a layer of burnt material covering the floor of the hollow. Two fragments of ground stone, a large flat stone, pieces of red ochre, and five articulated vertebrae from the spine of a wild aurochs had been left on the floor before the structure was burned. On top of the burned remains three caches of pierced marine shells had been placed, totalling more than a thousand beads in all, some from the distant Mediterranean, and the rest from the more distant Red Sea. Finally, archaeologically sterile sand was brought in some quantity to cover the visible traces of the burnt structure. Muheisen's excavations had found a small part of this sunken structure, and he found two male bodies buried below its floor; again, the association presages the close relation between burials and houses that is a feature of the Pre-Pottery Neolithic. The second structure is only partly excavated. Like the first, it was formed in an excavated depression, but it was not burned at the end of its use. The floor inside the structure showed signs of having been remade two or three times. Associated with its abandonment was a cache of gazelle and aurochs horns, another feature that recurs in the late Epipalaeolithic, and that has attracted attention and discussion in Pre-Pottery Neolithic contexts. Lisa Maher has been emphatic in her efforts to demonstrate that symbolic practices and attitudes that have long been considered characteristics of the Pre-Pottery Neo-lithic were clearly present in earlier forms not only in the late Epipalaeolithic (as we shall see in the next chapter) but equally from an early stage in the early Epipal-aeolithic (Macdonald et al. 2018; Maher 2018)s: in important ways, the Neolithic was not new.

Some distance to the south of Kharaneh IV is another large aggregation site, Jilat 6, which also dates to the early Epipalaeolithic. In the 1980s Andrew Garrard directed a multi-disciplinary project that surveyed and excavated in the Azraq basin in east-central Jordan. Garrard and his team put a great deal of work into understanding the palaeo-environmental history of the landscape worked un-der the changing climatic conditions of the final Pleistocene and early Holocene (Garrard 1998). One of the three areas in which they concentrated their survey work, small-scale soundings on a number of sites of different types, and larger exposures on a few sites was the dry valley of the Wadi Jilat. At this stage we are particularly interested in the Upper Palaeolithic and earlier Epipalaeolithic (Byrd 1990; Byrd et al. 2016; Byrd & Garrard 1994, 2017; Garrard 1998).

There are a few, very small Upper Palaeolithic sites in the Azraq project area. Three sites are radiocarbon dated to the early Epipalaeolithic period. There are a few heavy grinding implements of black basalt (which must have been brought some 50 km from the nearest source), but only scant traces of carbonised plant remains. One site in Wadi Jilat, Jilat 6, radiocarbon dated around 17,000 years ago, close to the end of the early Epipalaeolithic period, did produce some plant remains. There were seeds of various grasses and other plants characteristic of a steppe environment, and seeds of sedges, which would suggest that there were wetter patches in the wadi bed nearby (Colledge & Hillman 1988). In the dry and treeless environment of the steppe, the animal species hunted were predominantly gazelle, backed up by a wild equid species (an extinct species of the horse family, looking like a wild ass, but the size of zebra). There were perforated marine shell ornaments on all the sites, some from the Mediterranean (which is 200 km distant) and some from the Red Sea (which is 300 km away). These groups participated in extensive social exchange networks (Richter et al. 2011), that must have extended north-south throughout these marginal regions right down to the head of the Red Sea, as well as east-west across the barrier of Lake Lisan in the rift valley, linking groups in the steppe zone with those living in the Mediterranean woodland zone further west.

Over time, from the Upper Palaeolithic, through the Kebaran phase in the Last Glacial Maximum and on through the Geometric Kebaran in the Bølling-Allerød interstadial Byrd and Garrard (1994: 91) have noted that the average thickness of the archaeological deposits increases, and the sites cover greater areas. The site of Jilat 6, however, is exceptional both for its size – more than 18,000 m^2 – and the accumulated depth of occupation deposit (Garrard & Byrd 2013). It is intriguing to note that the chipped stone industry of Jilat 6 is a variant of the Geometric Kebaran;

while the occupations of Kharaneh IV and the more recently founded Jilat 6 partly overlap in time, the two sites were used by culturally and socially different groups. For both sites, Byrd et al. (2016) have modelled the foraging ranges of their hunter-gatherer occupants, and have concluded that the groups who came together there probably spent other parts of their year in small bands, either to the east in the wetlands of the Azraq basin itself, or to the west in the narrow woodland zone along the edge of the plateau above the rift valley. Both Kharaneh IV and Jilat 6 are situated in the semi-arid steppe, rather than within the Azraq basin itself, with its lake and wetlands. Reconstructing the extent of the foraging ranges, Byrd et al. (2016) conclude that almost all of the territory that was within a one- or two-day foraging range consisted of dry steppe. Clearly, there must have been water in the wadi beds at the seasons when both sites were occupied; and the main food resource would have been the seasonally large herds of gazelle, supported by wild equids, and a range of smaller prey, including hares, wolf, fox, ostrich and a variety of larger birds such as partridge. Sites close to the seasonal lake and springs of Azraq, which may have been hunting camps at a distance of one or two days walking, have also produced bones of wild cattle and of seasonally migrant waterfowl.

Before we leave the north of Jordan, the site of Uyun al-Hammam is certainly worth bringing into the picture of the early Epipalaeolithic. While there was a small but intensive occupation there dating to the early Epipalaeolithic, the site is remarkable for its cemetery of ritualised burials, which seem to have followed when the site was no longer used as a settlement (Maher 2005; Maher et al. 2011). Maher presses the point that there are early Epipalaeolithic precursors of practices that have commonly been associated with the Pre-Pottery Neolithic, but of course there are earlier examples of deliberate and ritualised burials associated with places of occupation, dating back to both Homo sapiens and Neanderthal burials in the later Middle Palaeolithic period. These early Epipalaeolithic burials associated with huts, houses, or settlements represent the amplification of earlier, if rarer, practices. At 'Uyun al-Hammam' the excavators found eight graves that had been dug into the existing Epipalaeolithic occupation deposit. Two were burials of single individuals, while two graves contained secondary burials (human remains gathered from elsewhere and reburied), and three more graves contained a second burial that had been added to an existing burial. Most of the bodies had been accompanied by grave goods, including chipped stone tools or weapons, ground stone implements, pieces of red ochre, and partial animal skeletons. Because it was not possible to identify precisely from which level the digging of any grave had begun (and that is a quite normal problem on excavation), it is not clear whether the burials were made while the site was in use as a hunter-gatherer base-camp, or whether it was an occupation site that later became a cemetery. One burial had been accompanied by the skull and a single leg-bone of a fox, while another grave contained a fox skeleton lacking its skull and a leg-bone. The excavators conclude that the two burials must have been made at the same time, and the remains of a fox was shared between them. There was also evidence of complicated rituals that involved the reopening of graves, the removal of body parts, and, in other graves, the addition of a skull or other body parts from elsewhere (Maher et al. 2011: 7). As we shall see over the following chapters, these early Epipalaeolithic burials are early examples of a remarkable continuity of ritual traditions stretching through the Epipalaeolithic and the Pre-Pottery Neolithic.

Turning now to the semi-arid south of Jordan, where there is nothing like the Azraq basin, there were different innovations in subsistence and settlement strategies. The key landscape feature is the rift valley, where the Dead Sea represents the southern end of the former Pleistocene Lake Lisan, and the Wadi Arabah continues the rift down to the head of the Gulf of Aqaba and the Red Sea. The landscape on either side of the rift valley above the Dead Sea in late Pleistocene times was open Mediterranean woodland, but further south, above the Wadi Arabah, the environment was semi-arid steppe and desert. On the west side, this is known as the Negev. The east side of the Wadi Arabah, the Jordanian side, is a high limestone plateau, with a series of deeply cut wadis leading down to the wide floor of the Wadi Arabah.

The settlement and subsistence strategies of the early Epipalaeolithic groups of southern Jordan have been reconstructed by Don Henry, who worked for many years on the Epipalaeolithic of southern Jordan. In Henry's reconstruction the early Epipalaeolithic hunter-gatherer groups divided their year between low altitude sites in the rift valley and the lower part of the wadis leading into it, and high-altitude sites on the edge of the plateau (Henry 1989: 167–70, 1995). This is essentially a novel strategy of seasonal transhumance, for which there is no evidence in the preceding Upper Palaeolithic period. The sites at low elevations tend to be larger and to have an appreciable depth of accumulated cultural deposits with high densities of a wide range of tools and equipment. The sites at high altitude tend to be smaller, more ephemeral, and to have low artefact densities. Henry argues that the low altitude sites were winter camps, where several small forager bands came together during the autumn and winter; the environment on the edge of the upland plateau would then have supported small, highly mobile forager bands during spring and early summer.

Across the rift valley, the Wadi Arabah, Nigel Goring-Morris undertook extensive survey, supported by small-scale excavations, on the Epipalaeolithic sequence in the Negev of southern Israel and the Sinai peninsula (Goring-Morris 1987). In the northern Negev, around the modern city of Beersheva, there is rainfall of almost 200 mm per year, but the land further south is progressively drier. Much of Sinai has rainfall of only around 50 mm per year. These are landscapes described as dry steppe, semi-desert and desert. There are indications of more moist conditions in the distant past in the form of relic trees, for example juniper, in various locations and wood charcoals from some archaeological sites, but most of the small trees and larger shrubs today are confined to the beds of the seasonal water-courses. Goring-Morris found few Kebaran sites of early Epipalaeolithic date, which would have dated to the Last Glacial Maximum. These were confined to locations around the edges of the highlands, and clustered around water sources. If this landscape is dry and harsh today, it must have been very difficult for hunter-gatherers during the Last Glacial Maximum, when conditions were notably drier. The Geometric Kebaran and Mushabian sites that represent the following phase of the Epipalaeolithic during the climatic amelioration of the Bølling-Allerød are much more numerous, and some of them are found at higher elevations.

The end of the middle and the beginning of the late Epipalaeolithic is the period with the greatest numbers of sites, and they are extensively distributed across the elevational range. Most of the sites are small, single-period occupations around a single hearth, but there are also more extensive sites with multiple hearths. Goring-Morris notes that sites in the highlands tend to have the greatest diversity of tool types (suggesting longer-term use as base camps) while sites in the lower elevations in the dunefields are slighter, have lower diversity of tool types, and are dominated by points (suggesting relatively high mobility and dependence on hunting). Like Henry, Goring-Morris has suggested that the different sizes and elevations of the sites represent seasonal transhumance and a degree of fission-fusion aggregation. He is hampered by the lack of bone or archaeo-botanical remains from these shallowly stratified or heavily deflated desert sites, and he can speculate on the seasonal occupation of the different zones only by reference to the likely climatic conditions, the availability of water, and the presumed distribution of plant foods. Like Henry with his Jordanian sites, Goring-Morris found no direct information on plant foods and almost no surviving animal bone. Goring-Morris therefore argues that the longer occupation and more general base camps in the highlands would have been occupied in the winter, when water sources would have been more generalised and the population would not have been 'tethered' to the lowland permanent water sources. On the other hand, he refers to the highlands as being likely the best sources of nuts, fruits and seeds, which would have been available in the summer and into the late summer. These nutritious harvests might have maintained the population through the autumn and into the winter. Perhaps the transhumance cycle was not a simple summer-winter alternation, but rather a summer to early winter season in the highlands, followed by a late winter and spring season at lower altitudes.

There has been much less research on the Epipalaeolithic or earlier Palaeolithic in the north Levant, both in the west of Syria and southeast Turkey. There are some Upper Palaeolithic and Epipalaeolithic sites in the el-Kowm depression, a basin of inland drainage about 100 km north-east of the desert city of Palmyra and 75 km south of the Euphrates valley. The surrounding landscape is dry steppe or desert, but the el-Kowm depression has a number of artesian springs and there is also a shallow, seasonal lake. The ground-water sources make the area an oasis. Epipalaeolithic sites outnumber those of the Upper Palaeolithic, but the sites in this arid region are severely deflated, leaving only stone tools scattered on the surface. There is a lack of charcoal or carbonised plant material for radiocarbon dating, which means that the only means of dating these sites is by comparison of the chipped stone assemblages with the established and dated cultural sequence of the southern Levant. Marie-Claire Cauvin (1990) has reported the existence of sites with a chipped stone industry resembling the Geometric Kebaran.

Summary

The early Epipalaeolithic period is of critical importance for understanding the processes that initiated the Epipalaeolithic–Neolithic transformation. While we have the relatively rich information from the Mediterranean woodland and adjoining semi-arid zones of the southern Levant, we have almost nothing from other parts of southwest Asia. This asymmetry in our knowledge is a consequence of the relative intensity of research in Israel and Jordan over many years by contrast with the lack of comparable fieldwork elsewhere. The problem is compounded because finding Epipalaeolithic sites requires specialist survey programmes. Most survey work in Southwest Asia has been carried out by teams searching the landscape for tell-sites, that is, settlement mounds, which represent the standard settlement type from the Neolithic onwards through the emergence of urbanism and the period of the ancient civilisations of the Near East; searching for Palaeolithic sites involves looking for caves and rock-shelters in the slopes of mountains, hill-sides and the sides of wadis, or recognising chipped stone scatters in open country. In the valleys of the Zagros mountains along the Iraq-Iran border there were surveys and small-scale excavations in the 1950s and 1960s. Middle Palaeolithic and Upper Palaeolithic sites were found in caves and rock-shelters. Epipalaeolithic layers overlay the Upper Palaeolithic in some sites, but there seems to be a significant break in occupation during the Last Glacial Maximum; where we have radiocarbon dates, the Epipalaeolithic occupations belong in the late Epipalaeolithic period (and will find their place in the following chapter).

Within the southern Levant we can see that population density increased significantly in the early Epipalaeolithic; there are more sites dating to the early Epipalaeolithic, despite the fact that the period is less than half the duration of the Upper Palaeolithic. Hunter-gatherer groups were beginning to operate differently from their Middle Palaeolithic and Upper Palaeolithic predecessors. Rather than being seriously constrained by the climatic effect of the LGM, some communities began to broaden the range of foods that they collected to include nuts, fruits and seeds of cereals, grasses and legumes. More importantly, they began to concentrate on harvesting seeds and grains that could be stored. They also began to establish seasonal camps, or even regular year-round occupation sites, where they could store these new foods and consume them over a period or, in a few places, even year-round. As Maher has pointed out, from her investigations at Uyun al-Hammam and Kharaneh IV, many of the characteristics that have been thought to distinguish the Pre-Pottery Neolithic or the late Epipalaeolithic are now known to have been present in the early Epipalaeolithic (Maher 2018). We have seen the complex use of space and buildings, 'place-making', engagement in long-distance exchange networks, a trend to transhumance and seasonal aggregation, in a few cases approaching sedentism; and there are increasing examples of elaborate symbolic behaviour such as ritualised human burial associated with the built environment.

References

Belfer-Cohen, A., & Goring-Morris, A. N. (2017). The Upper Palaeolithic in Cisjordan. In Y. Enzel & O. Bar-Yosef (Eds.), *Quaternary of the Levant: Environments, Climate Change, and Humans.* pp. 627–638. Cambridge: Cambridge University Press.

Belfer-Cohen, A., & Goring-Morris, N. (2018). An Anthropological Review of the Upper Paleolithic in the Southern Levant: From Prehistory to the Present. In A. Yasur-Landau, E. H. Cline, & Y. M. Rowan (Eds.), *The Social Archaeology of the Levant* (pp. 29–46). Cambridge: Cambridge University Press.

Byrd, B. (1990). Late Pleistocene Settlement Diversity in the Azraq Basin. In O. Aurenche, M.-C. Cauvin, & P. Sanlaville (Eds.), *Préhistoire du Levant: processus des changements culturels; hommage à Francis Hours* (pp. 257–64). Paris: Éditions du CNRS.

Byrd, B., & Garrard, A. (1994). The Last Glacial Maximum in the Jordanian Desert. In C. Gamble & O. Soffer (Eds.), *The World at 18,000 BP, Volume Two: Vol. 2, Low Latitudes* (pp. 78–96). London: Unwin Hyman.

Byrd, B. F., & Garrard, A. N. (2017). The Upper and Epipalaeolithic of the Azraq Basin, Jordan. In Y. Enzel & O. Bar-Yosef (Eds.), *Quaternary of the Levant: Environments, Climate Change, and Humans* (pp. 669–78). Cambridge: Cambridge University Press.

Byrd, B. F., Garrard, A. N., & Brandy, P. (2016). Modeling Foraging Ranges and Spatial Organization of Late Pleistocene Hunter–Gatherers in the Southern Levant – A Least-Cost GIS Approach. *Quaternary International*, 396, 62–78. doi:10.1016/j.quaint.2015.07.048

Cauvin, M.-C. (1990). L'oasis d'el Kowm et le Kébarien Géométrique. In O. Aurenche, M.-C. Cauvin, & P. Sanlaville (Eds.), *Préhistoire du Levant: processus des changements culturels* (pp. 270–82). Paris: Éditions du CNRS-Paléorient.

Colledge, S., & Hillman, G. (1988). The Plant Remains, In Garrard, A. N. & Hillman, G. Environment and Subsistence during the Late Pleistocene and Early Holocene in the Azraq Basin. *Paléorient*, 14, 40–8.

Garrard, A. (1998). Environment and Culture Adaptations in the Azraq Basin: 24,000–7000 BP. In D. O. Henry (Ed.), *The Prehistoric Archaeology of Jordan* (pp. 139–48). Oxford: Archaeopress.

Garrard, A. N., & Byrd, B. F. (2013). *Beyond the Fertile Crescent: Late Palaeolithic and Neolithic Communities of the Jordanian Steppe. The Azraq Basin Project, Volume 1: Project Background and the Late Palaeolithic (Geological Context and Technology).* Oxford: Oxbow Books.

Garrod, D. A. E., Bate, D. M. A., McCown, T. D., & Keith, A. (1937). *The Stone Age of Mount Carmel. Joint Expedition of the British School of Archaeology in Jerusalem and the American School of Prehistoric Research (1929–1934)* (Vol. 2 vols.). Oxford: The Clarendon Press.

Goring-Morris, A. N. (1987). *At the Edge: Terminal Pleistocene Hunter-Gatherers in the Negev and Sinai.* Oxford: BAR International Series 361.

Goring-Morris, A. N., & Belfer-Cohen, A. (2017). The Early and Middle Epipalaeolithic of Cisjordan. In Y. Enzel & O. Bar-Yosef (Eds.), *Quaternary of the Levant: Environments, Climate Change, and Humans* (pp. 639–50). Cambridge: Cambridge University Press.

Henry, D. O. (1989). *From Foraging to Agriculture: The Levant at the End of the Ice Age.* Philadelphia: University of Pennsylvania Press.

Henry, D. O. (1995). *Prehistoric Cultural Ecology and Evolution: Insights from Southern Jordan.* New York, NY; London: Plenum Press.

Hershkovitz, I., Marder, O., Ayalon, A., Bar-Matthews, M., Yasur, G., Boaretto, E., . . . Barzilai, O. (2015). Levantine Cranium from Manot Cave (Israel) Foreshadows the First European Modern Humans. *Nature*, 520(7546), 216–9. doi:10.1038/nature14134

Hershkovitz, I., Speirs, M. S., Frayer, D., Nadel, D., Wish-Baratz, S., & Arensburg, B. (1995). Ohalo II H2: A 19,000-Year-Old Skeleton from a Water-Logged Site at the Sea of Galilee, Israel. *American Journal of Physical Anthropology*, 96(3), 215–34. Retrieved from https://doi.org/10.1002/ajpa.1330960302

Hodder, I. (1990). *The Domestication of Europe: Structure and Contingency in Neolithic Societies*. Oxford: Basil Blackwell.

Kaufman, D. (1989). Observations on the Geometric Kebaran: A View from Neve David. In O. Bar-Yosef & B. Vandermeersch (Eds.), *Investigations in South Levantine Prehistory* (pp. 275–86). Oxford: British Archaeological Reports, International Series.

Kaufman, D., & Ronen, A. (1987). La sépulture Kébarienne Géométrique de Névé David, Haifa, Israël. *L'Anthropologie*, 91, 335–42.

Kuhlwilm, M., Gronau, I., Hubisz, M. J., de Filippo, C., Prado-Martinez, J., Kircher, M., . . . Castellano, S. (2016). Ancient Gene Flow from Early Modern Humans into Eastern Neanderthals. *Nature*, 530, 429–33. doi:10.1038/nature16544

Liu, C., Shimelmitz, R., Friesem, D. E., Yeshurun, R., & Nadel, D. (2020). Diachronic Trends in Occupation Intensity of the Epipaleolithic Site of Neve David (Mount Carmel, Israel): A Lithic Perspective. *Journal of Anthropological Archaeology*, 60, 101223. doi:10.1016/j.jaa.2020.101223

Macdonald, D. A., Allentuck, A., & Maher, L. A. (2018). Technological Change and Economy in the Epipalaeolithic: Assessing the Shift from Early to Middle Epipalaeolithic at Kharaneh IV. *Journal of Field Archaeology*, 43(6), 437–56. doi:10.1080/00934690.2018 .1504542

Macdonald, D. A., & Maher, L. A. (2022). Leaving Home: Technological and Landscape Knowledge as Resilience at Pre-Holocene Kharaneh IV, Azraq Basin, Jordan. *Holocene*, 32(12), 1450–61. doi:10.1177/09596836221121784

Maher, L. (2005). Recent Excavations at the Middle Epipalaeolithic Encampment of Uyyun al-Hammam, Northern Jordan. *Annual of the Department of Antiquities of Jordan*, 49, 101–14.

Maher, L. A. (2018). Persistent Place-Making in Prehistory: The Creation, Maintenance, and Transformation of an Epipalaeolithic Landscape. *Journal of Archaeological Method and Theory*. doi:10.1007/s10816-018-9403-1

Maher, L. A., Allentuck, A., Martin, L., Spyrou, A., & Jones, M. D. (2016). Occupying Wide Open Spaces? Late Pleistocene Hunter–Gatherer Activities in the Eastern Levant. *Quaternary International*, 396, 79–94. doi:10.1016/j.quaint.2015.07.054

Maher, L. A., Richter, T., & Stock, J. T. (2012). The Pre-Natufian Epipaleolithic: Long-Term Behavioral Trends in the Levant. *Evolutionary Anthropology*, 21(2), 69–81. doi:10.1002/ evan.21307

Maher, L. A., Stock, J. T., Finney, S., Heywood, J. J. N., Miracle, P. T., & Banning, E. B. (2011). A Unique Human-Fox Burial from a Pre-Natufian Cemetery in the Levant (Jordan). *PLoS ONE*, 6(1), e15815. Retrieved from https://doi.org/10.1371%2Fjournal. pone.0015815

Muheisen, M. (1988). The Epi-Palaeolithic Phases of Kharaneh IV. In A. N. Garrard & H. G. K. Gebel (Eds.), *The Prehistory of Jordan: The State of Research in 1986* (pp. 353–67). Oxford: British Archaeological Reports International Series 396.

Nadel, D. (2017). Ohalo II: A 23,000-Year-Old Fisher-Hunter-Gatherer's Camp on the Shore of Fluctuating Lake Kinneret (Sea of Galilee). In O. Bar-Yosef & Y. Enzel (Eds.),

Quaternary Environments, Climate Change, and Humans in the Levant (pp. 291–4). Cambridge: Cambridge University Press.

Nadel, D., Piperno, D. R., Holst, I., Snir, A., & Weiss, E. (2012). New Evidence for the Processing of Wild Cereal Grains at Ohalo II, a 23,000-Year-Old Campsite on the Shore of the Sea of Galilee, Israel. *Antiquity*, 86(334), 990–1003.

Nadel, D., Weiss, E., Simchoni, O., Tsatskin, A., Danin, A., & Kislev, M. (2004). Stone Age Hut in Israel Yields World's Oldest Evidence of Bedding. *Proceedings of the National Academy of Sciences of the United States of America*, 101(17), 6821–6. doi:10.1073/pnas.0308557101|ISSN 0027-8424

Nadel, D., & Werker, E. (1999). The Oldest Ever Brush Hut Plant Remains from Ohalo II, Jordan Valley, Israel (19,000 BP). *Antiquity*, 73, 755–64.

Neuville, R. (1934). Le prehistorique de Palestine. *Revue Biblique*, 43(2), 237–59.

Piperno, D. R., Weiss, E., Holst, I., & Nadel, D. (2004). Starch Grains on a Ground Stone Implement Document Upper Paleolithic Wild Cereal Processing at Ohalo II, Israel. *Nature*, 430, 670–3. doi:10.1038/nature02734

Richter, T., Garrard, A. N., Allock, S., & Maher, L. A. (2011). Interaction before Agriculture: Exchanging Material and Sharing Knowledge in the Final Pleistocene Levant. *Cambridge Archaeological Journal*, 21(1), 95–114. doi:10.1017/s0959774311000060

Snir, A., Nadel, D., Groman-Yaroslavski, I., Melamed, Y., Sternberg, M., Bar-Yosef, O., & Weiss, E. (2015a). The Origin of Cultivation and Proto-Weeds, Long before Neolithic Farming. *PLoS ONE*, 10(7), e0131422. doi:10.1371%2Fjournal.pone.0131422

Snir, A., Nadel, D., & Weiss, E. (2015b). Plant-Food Preparation on Two Consecutive Floors at Upper Paleolithic Ohalo II, Israel. *Journal of Archaeological Science*, 53, 61–71.

Tillier, A.-M., & Arensburg, B. (2017). Neanderthals and Modern Humans in the Levant: An Overview. In Y. Enzel & O. Bar-Yosef (Eds.), *Quaternary of the Levant: Environments, Climate Change, and Humans* (pp. 607–10). Cambridge: Cambridge University Press.

Villanea, F. A., & Schraiber, J. G. (2019). Multiple Episodes of Interbreeding between Neanderthal and Modern Humans. *Nature Ecology & Evolution*, 3(1), 39–44. doi:10.1038/s41559-018-0735-8

Watkins, T. (1990). The Origins of House and Home? *World Archaeology*, 21(3), 336–47.

Weiss, E. (2009). Glimpsing into a Hut: The Economy and Society of Ohalo II's Inhabitants. In A. Fairbairn & E. Weiss (Eds.), *From Foragers to Farmers: Papers in Honour of Gordon C. Hillman* (pp. 153–60). Oxford: Oxbow Books.

Weiss, E., Kislev, M. E., Simchoni, O., Nadel, D., & Tschauner, H. (2008). Plant-Food Preparation Area on an Upper Paleolithic Brush Hut Floor at Ohalo II, Israel. *Journal of Archaeological Science*, 35(8), 2400–14. doi:10.1016/j.jas.2008.03.012

Weiss, E., Wetterstrom, W., Nadel, D., & Bar-Yosef, O. (2004). The broad spectrum revisited: Evidence from plant remains. *Proceedings of the National Academy of Sciences of the United States of America*, 101(26), 9551–9555. doi:10.1073/pnas.040362101|ISSN 0027-8424

Wright, K. (1991). The Origins and Development of Ground Stone Assemblages in Late Pleistocene Southwest Asia. *Paléorient*, 17(1), 19–45.

5 Complex Hunter-Harvesters in the Levant and beyond

The division of the Epipalaeolithic period between two chapters is necessary because of the material that needs to be covered; there is so much important information from the late Epipalaeolithic that it requires a chapter of its own. The late Epipalaeolithic in the Levant, especially the southern Levant, usually labelled the Natufian, has been the focus of a disproportionate amount of research, discussion and debate. The Natufian has provided enough material to fill two conference volumes that together count more than 1350 pages (Bar-Yosef & Valla 1991, 2013).

Having said that, there is more to the late Epipalaeolithic than some detailed changes in the chipped stone industry; as we shall see, there was something of a cultural step-change between the early Epipalaeolithic and the Natufian. The late Epipalaeolithic in the southern Levant begins during the Bølling-Allerød interstadial warming after the Last Glacial Maximum, around 14,000 or 13,000 BCE, and continues through the Younger Dryas cold phase, meeting the following early Pre-Pottery Neolithic period at the beginning of the Holocene, around 9600 BCE. The late Epipalaeolithic period is represented beyond the southern Levant in other parts of the hilly flanks zone, and in central and southern Anatolia. This chapter will therefore have two (unequal) parts: we start with the Natufian in the southern Levant, and in the second part of the chapter we review the evidence for the late Epipalaeolithic elsewhere in our region.

The Natufian

Those who know best the archaeology of the Epipalaeolithic period in Syria, Israel and Jordan, scholars like Ofer Bar-Yosef and François Valla, who arranged the conferences that produced the two volumes just mentioned, and Don Henry (Henry 1995: 328), are agreed that the last phase of the Epipalaeolithic period is sharply differentiated from the earlier phases. Indeed, Don Henry (1989: 180) has argued that the Natufian represents a 'pre-agricultural revolution', a prelude to the Neolithic. While archaeologists sometimes refer to the earlier Palaeolithic phenomena as 'cultures', they are usually reliant on the chipped stone industries: when we reach the Natufian phenomenon, we can for the first time recognise a complex of material traits and cultural practices within specific geographical and chronological constraints that can appear to conform with Gordon Childe's classic definition

DOI: 10.4324/9781351069281-6

of a cultural group. I immediately want to add, however, that one of the cultural characteristics of the Natufian is the way that different Natufian communities were adapted to the specifics of their local environment (Belfer-Cohen & Goring-Morris 2013). There are also many more sites of the Natufian period known than of the earlier Epipalaeolithic period. These observations indicate important points about this latter stage in the Epipalaeolithic period. However, given what we now know of the earlier Epipalaeolithic, with sites like Ohalo II, Jilat 6, Uyun al-Hammam, and Kharaneh IV that we saw in the previous chapter, the extent to which the Natufian seems to represent a step-change is now more nuanced.

The Natufian chipped stone industry is closely related to its Epipalaeolithic predecessors; there are relatively small changes from earlier traditions. François Valla (1984) has made a definitive study of the Natufian chipped stone industry, and Don Henry (1989: 185–95) offers a general discussion of other cultural traits. In fact most of what is characteristic of the Natufian has clear antecedents in the early Epipalaeolithic: the phenomenon that has impressed the specialists is essentially one of intensification. If there were hints of sedentary or semi-sedentary settlements earlier in the Epipalaeolithic, permanent settlements clearly existed in the Natufian. If some occupation sites of the earlier Epipalaeolithic (setting aside the very special aggregation sites in Jordan) were larger than earlier Palaeolithic occupations, there are distinctly larger settlements in the Natufian. If people in earlier periods used some ground stone implements for pounding and grinding, in the Natufian the increased numbers of these heavy pieces of equipment cannot help but impress (15.2% of Geometric Kebaran sites have produced ground stone implements, but almost 50% of Natufian sites). If there are a few reports of deliberate human burials at early Epipalaeolithic sites, there is an impressive number of cemeteries of burials, and burials with ceremony, ritual and grave goods at some Natufian sites.

The third and fourth points mentioned, those concerning the range of the material culture repertoire and the number of sites of the Natufian, relate to different aspects of the same phenomenon. A number of Natufian communities were more sedentary than their predecessors; at a number of locations Natufian communities lived year-round, and stayed there over long periods, living together in relatively large groups. Archaeologists have sought to demonstrate that there is a general correlation between the range of discarded tools and the range of debris on the one hand and the range of activities that were pursued at a site on the other hand. On this basis base camps where groups lived for longer periods and carried out the whole range of their activities can be distinguished from single-activity, short-term occupation camps. Similarly, ethnographers have observed that people who live in small, mobile hunter-gatherer bands have very few possessions and a minimum of equipment, whereas for sedentary people, like ourselves, there is no need for concern about developing particular tools for particular tasks, or owning large, heavy pieces of non-portable equipment. Thus, we may see the fuller range of material culture among the Natufians as being correlated with their more sedentary settlements and the full range of activities that were carried out within them.

The Natufian in the southern Levant covers a period of at least 3000 years, during which time there were cultural changes. On the basis of very small changes in the chipped stone tool industry observed in her excavations at the cave of el Wad in the Mt Carmel hills Dorothy Garrod distinguished an early from a late Natufian (Garrod et al. 1937). And, since that time, a third stage has been recognised, which has been labelled the final Natufian. The Natufian, therefore, changed over time: three stages within 3000 years. We have reached a stage in our story when detectable cultural change was happening at a faster tempo.

At one time, archaeologists thought that the early Natufian was a boom period, while the fewer sites of the late and final Natufian, and their less impressive and substantial archaeological remains, was thought to indicate a reduction in population and a return to mobile foraging. When we learned to the existence of the colder, drier Younger Dryas phase, it was suggested that the late and final Natufian showed the effects of that environmental deterioration. Such a view has been shown to be false, because, by chance, many more sites of late and final Natufian age have come to light, and some of them are indeed impressive and substantial. A careful examination of the radiocarbon chronologies of the Younger Dryas phase and the phasing of the Natufian shows that the cultural change that marks the transition from the early to the late Natufian occurred during the Bølling-Allerød amelioration and many centuries before the Younger Dryas began. And the archaeological evidence now shows that the final Natufian represents an expansion in population numbers and geographical range (Henry 2013).

One of the most awkward characteristics of the Natufian for archaeologists is the lack of uniformity: whatever one chooses as a distinctive cultural characteristic, for example, circular stone-built houses, or cemeteries of ritualised burials, turns out to be present at some sites, but not at most others. The variety over time and across the region has been discussed by Anna Belfer-Cohen and Nigel Goring-Morris (2013), two senior Israeli archaeologists who have been working on the Natufian since they began doctoral research. Each site is different from any other in one or several regards. The solution here is to focus on some of the features of Natufian sites that are novel or impressive, so that we can see, by comparing this chapter with its predecessor, how the late Epipalaeolithic Natufian indeed represents something of a step-change from the earlier Epipalaeolithic, impressive as that was.

Eynan ('Ain Mallaha)

There is one site Natufian site that requires a few paragraphs of its own. It has been carefully investigated in a series of episodes of excavation and research over more than 70 years. Over that long period of research, the excavators have substantially changed their understanding of the settlement's history. Eynan has come to play a disproportionate role in the development of an understanding of the late Epipalaeolithic in the southern Levant as a whole. Eynan ('Ain Mallaha, its Arabic name), occupies a slope that leads down to the narrow strip of level ground surrounding Lake Huleh in the far north of Israel. When the site was occupied, between about 13,000 and 10,000 BCE, its inhabitants would have looked across the shallow waters of the lake, but over the millennia it has silted up. In modern times, the has been drained and the site was discovered when water management works were being undertaken in the mid-1950s.

The early accounts of the work at Eynan described a settlement that was a powerful presence in the early Natufian, but that seemed to have become ephemeral and only occasionally occupied in the late and final Natufian. This chimed with the interpretation of the pollen core from the bed of Lake Huleh, which was interpreted as recording the serious impact of the Younger Dryas climatic reversal on the tree species of the Mediterranean woodland zone. As we saw in Chapter 2, the impact of the Younger Dryas was quite varied and generally less dramatic on the different landscapes of Southwest Asia: and the more recent excavations at Eynan have shown that the site continued to be occupied throughout the late and final Natufian, although the architecture was on a smaller scale and less substantial. We should not be too influenced by the less impressive buildings of the late phase of the settlement, for Laure Dubreuil (2004) found evidence of increasing use of ground stone equipment for pounding and grinding cereals in the final Natufian, suggesting that, at least in the Natufian core area, storable plant food resources were used more intensively than before.

Research on the material from the Eynan excavations has produced book-length studies: there are monographs on the chipped stone industry (Valla 1984), the faunal remains (Bouchud 1987), and the burials and physical anthropology of the human remains (Perrot et al. 1988). Here I rely heavily on a recent summary by François Valla, who has been the research project's director for more than 20 years, and his colleagues (Valla et al. 2017). The first inhabitants chose a prime location for their settlement; it was a place from which a variety of complementary ecological zones could be accessed (Figure 5.1). There was a rich variety of resources to be had in the surrounding hills, on the low ground around Lake Huleh, in its shallow margin and reedbeds, and its deeper waters. The inhabitants of Eynan hunted gazelle, fallow deer, wild boar, red and roe deer, hare, tortoise, reptiles and amphibians, and the migrant waterfowl that visited the lake. They fished the lake, and gathered terebinth and almonds, small-seeded grasses and some wild barley and wheat.

The substantial remains of circular buildings were cut into the sloping ground with dry-stone masonry retaining walls and floors lower than the outside ground level. The buildings of the final phase seem to have been less substantial, perhaps because their floor levels were not cut so deeply into the ground, and they required less use of stone retaining walls. The final phase was only intensively investigated in the most recent excavations.

Houses usually possessed a central hearth, sometimes defined by stone slabs, and the floors were cluttered with ground stone implements – querns and grinding stones, mortars and pestles. For the earlier phase of occupation, the excavators found three large circular structures and part of a fourth, and about 30 burials. At first glance, many of the burials are associated with two of the buildings, numbers 1 and 131, but Valla et al. (2017) has carefully pointed out that many of the burials were already in place before the buildings were constructed, and the rest of the burials were interred after the structures had been abandoned and filled in. We have some detail concerning the unusual large building 131 (Valla 1988), which was D-shaped and about 8 metres in diameter (Figure 5.2). The use of the building was not simply domestic. It had three hearths, two of which were areas of burning confined by an arc of stones and situated between roof support posts at the rear

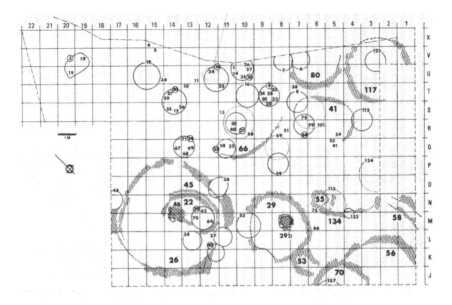

Figure 5.1 Plan of part of the settlement of Eynan. There was a cluster of circular houses with stone walls, and many cylindrical pits carefully lined with mud-plaster, some of which had been re-used for burials. (Valla 1991, Fig 3, with permission)

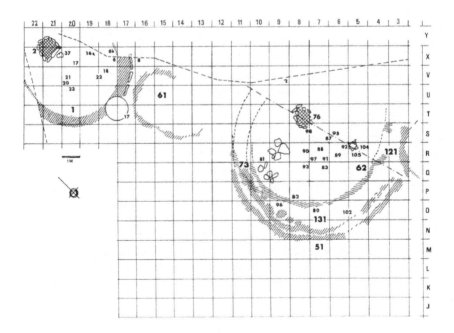

Figure 5.2 Large stone-built structures of the earliest phase at Eynan. Both Buildings 1 and 131 were associated with many burials. (Valla 1991, Fig 1, with permission)

of the building. Among the things that were left in the building some might be associated with the functions of everyday life, such as preparing flint cores, making tools and weapons, and preparing food. Other things had less obvious purposes: there were the polished and incised antler of a roe deer, two fragments of a piece of sculpted limestone, the vault of a human skull, half of a dog's jawbone, a heap of butchered gazelle bones and a collection of small pebbles. Brian Boyd (1995) has sought to build on Valla's detailed analyses and observations, emphasising the symbolic nature of much of the activity associated with the objects found in the structure, and its association most probably with the 'de-commissioning' of the building prior to its (deliberate) destruction. We need to remember that structure 131 was built directly above the location of a cluster of burials, and that another cluster of burials was made after its destruction. The most recent excavations have better documented the final Natufian phase (Samuelian 2013; Valla et al. 2013). The houses were smaller and of less substantial construction; there were extensive areas of cobbled yards, and some burials in simple pits (Figure 5.3).

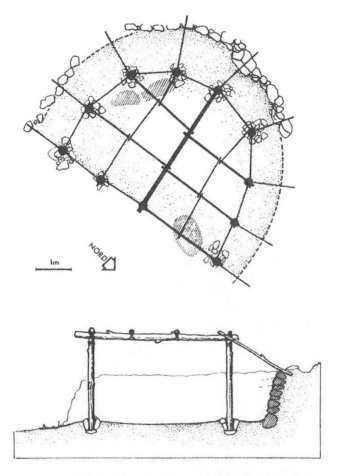

Figure 5.3 Reconstruction of the large, open-sided, D-shaped Building 131 at Eynan. (Valla 1991, Fig 2, with permission)

Large, open-air settlements are rare, but Eynan is partnered in some ways by the site of Wadi Hammeh 27 in Jordan. Like a number of sites that we shall be encountering, Wadi Hammeh 27 acquired its name when sites were located (and numbered) in the course of survey work along the length of a wadi or valley that drains down into the Jordan valley from the edge of the Jordanian highlands. Situated on a ridge in the wadi bed, the site has been seriously eroded; probably less than half of the original site of perhaps 5000 m^2 has survived. Wadi Hammeh 27 was occupied in the early Natufian phase, around 12,000 BCE (Figure 5.4).

Like the early phase at Eynan, the excavated area was filled by two large circular structures with stone walls. Both buildings had been rebuilt at least twice on the same location. One structure was horseshoe shaped, with a large opening at one side. The other, Structure 2, in its final phase consisted of three concentric rings of stone-built wall and a central pile of large limestone boulders. Three engraved stone slabs had been set on edge in the floor, and were placed end-to-end; one side of each had been carved with a concentric lozenge motif (Figure 5.5). The interiors of the two structures were filled with a strange variety of settings of stone and caches of different kinds; there was, for example, a pile of beautifully made black basalt mortars and pestles, a cluster of tools, among which was a sickle haft carved from a goat horn core, tools, and a cache of dentalium shells (from the Mediterranean shore). Analysis of the finds showed that most of the chipped stone and other debris was deposited inside the two structures. Perhaps the least expected finds were the

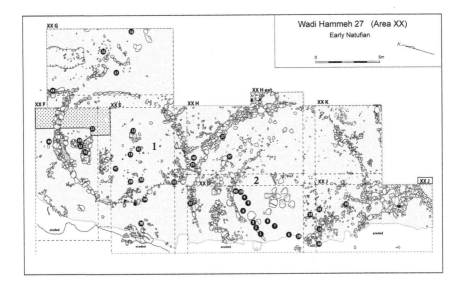

Figure 5.4 Wadi Hammeh 27, plan of the two stone-built semi-subterranean structures. The plan shows the many pieces or human crania that were concentrated within both buildings. Many other kinds of artefact were similarly concentrated within the two buildings. (Edwards 2009, Fig. 1, with permission)

many (mostly burnt) fragments of human skulls that were found scattered around the floors of both structures. Altogether, this is not the picture one could expect of ordinary daily life in ordinary everyday house; clearly we do not understand how these communities functioned and what their buildings meant to them.

The Natufian was first identified by Dorothy Garrod, who excavated over a period of six years beginning in 1929 in the cave-sites of the Mount Carmel hills in what is now northern Israel, (Garrod et al. 1937). The key site for the Natufian was the cave of el Wad. Since those first investigations there have been several more episodes of excavation at el Wad. There was occupation within the cave itself, but there was much more Natufian period occupation terraced into the slope in front of the cave. Natufian groups were larger than their Palaeolithic predecessors, but they still liked to focus their settlement on a cave. They also buried their dead in numbers at el Wad. Renewed excavations and analysis of the older excavation records has produced a detailed account of the complicated long-term use of the site (Weinstein-Evron et al. 2013). Following the first use of the site as a base-camp, it became a cemetery, with dozens of burials both in the cave and on the terrace outside the cave-mouth. The site was reoccupied as a settlement, of which substantial remains of walls, structures, and paved areas have been excavated. Finally, in the last centuries of the Natufian, the site once again became a cemetery. Like a number of burials elsewhere, several of the el Wad bodies were found with large numbers of a small, cylindrical marine shells (dentalium) on their skulls or around the shoulder area. The shells must have been sewn onto caps or capes of some organic material.

There is a nice little detail that enables us to be sure that there was permanent settlement at el Wad. Among the micro-fauna from the occupation debris the house mouse has been identified, and the house mouse only became an identifiable and

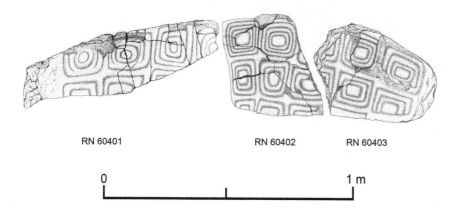

RN 60401   RN 60402   RN 60403

0   1 m

Figure 5.5 Wadi Hammeh 27. Three decorated slabs, originally a single large slab, were set on edge in a line in the larger circular structure in Figure 5.4. (By courtesy of P. Edwards)

separate species once there were permanent human occupations where it could become a new species (Tchernov & Valla 1997; Weinstein-Evron et al. 2007: 98; Weissbrod et al. 2017). The broad spectrum of faunal remains from the major period of occupation of the terrace outside the el Wad cave-mouth has also convinced the researchers that the site was permanently and intensively occupied in the long term (Yeshurun et al. 2014). Earlier researchers had interpreted the long-term decrease in the numbers of hunted large herbivores and the increase in the numbers of small animals, birds, and tortoises as evidence for resource depression in the face of increasing human population density in the Levant. An alternative reconstruction makes better sense: the increased reliance on medium-sized (primarily gazelle), small animals, birds and reptiles actually fits well with the adoption of sedentary life. In permanent settlements people need to rely on immediately local resources, and species that reproduce quickly and can tolerate intensive hunting and trapping.

While we have concentrated on two key sites that have produced burials, there are other Natufian sites where there are none. But there is one site, Hilazon Tachtit, which is unique (Figure 5.6). It is a small cave site in the Galilee region of northern Israel and it dates to the end of the Natufian period, around 10,000 BCE. It was a cemetery in which at least 28 individuals were buried, but one burial was

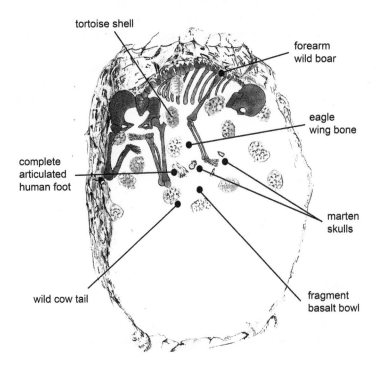

Figure 5.6 The extraordinary elaborate burial of a female at Hilazon Tachtit dates to the later part of the Late Epipalaeolithic. The plan has been simplified; some of the associated finds have been omitted. (Grosman et al. 2008, Fig. 4, by courtesy of the authors)

exceptional (Grosman et al. 2008; Grosman & Munro 2016; Munro & Grosman 2010). While there are also the kinds of remains – chipped stone tool debris, other tools, animal bone – that are found on occupation sites, there is not space for this site to have been used for regular occupation by a group; rather, the remains are probably the debris from the occasions when people were engaged in the events surrounding the burials. There was evidence of feasting events; two round or oval cavities excavated into the rocky floor, in which the bones of wild cattle and the shells and leg-bones of many tortoises had been carefully buried (Munro & Grosman 2010). The human remains had been crammed into three pits, and none were complete skeletons. Each of the pits had been used for successive burial events, sealed with flat stone slabs. The ceremonies also involved the re-opening of pits and the removal of, for example, the long bones of arms and legs, and the skulls. The buried bodies included both sexes and all ages (Grosman & Belfer-Cohen 2022).

The special burial had been placed at the base of one of two cavities excavated in the floor of the cave, and evidence of feasting had been piled into the upper part of the cavity. The roughly cut base and walls of the oval cavity were lined with mud plaster, and there was then a floor made of small slabs of limestone. Before the body was placed in the pit, a strange array of things was carefully placed on the floor slabs, including a complete horn core and part of the skull of a gazelle, three unusual marine shells, the broken base of a basalt palette, a lump of red ochre, a chalk fragment, and three tortoise shells. The body of a mature female was then placed in a semi-sitting position against the side of the cavity. She was accompanied by an extraordinary array of special things: the fore-leg of a wild boar, more than 80 tortoise shells, part of the skull of a stone marten, the tail of an aurochs, the wing tip of a golden eagle, a single bone of a leopard, two marine shell pendants, a bone point, an articulated human foot, and three stone tools. The fill above the burial was dense with animal bone, including large amounts of tortoise parts, and the bones of at least eight gazelles, of which four were newborns. Finally, a large flat capstone was placed to close the pit. The woman buried with such ceremony and such a strange array of things was clearly someone very special. The excavators suggest that the odd collection of things were the paraphernalia of her identity as a shaman (Grosman et al. 2008; Grosman & Munro 2016, with comments from a number of other archaeologists).

There are a very few, beautifully realised carvings of animals or parts of animals from Natufian sites; the best known are two antler sickle hafts, one from Kebara, the other from el Wad. A common feature of Natufian sites is the frequency of items of personal ornament. These are most obvious when they are found on skeletons in burials, and Natufian burials with dentalium shell ornaments have been known since Dorothy Garrod's excavations at el Wad (Garrod et al. 1937). The number of cemeteries has grown, and with it the number of occurrences of bodies wearing caps, hair ornaments, capes, bracelets and garters consisting of perforated dentalium shells threaded onto a bracelet or stitched onto some organic medium. There are shells of other species, most of them from the Mediterranean, but some from the Red Sea, as well as perforated animal teeth and stone beads and pendants. On sites where there have been no burials found occasional finds of marine shells have been reported with a frequency that contrasts greatly with the preceding millennia of the Epipalaeolithic period.

These little objects are also indicators of the more intensive operation of exchange networks. It was presumably exchange that brought the black basalt to Natufian settlements that are far from any geological source of the material. And it is also at this period that occurrences of central Anatolian obsidian (many hundreds of pieces from the final period of occupation at Eynan alone) indicate that the exchange network in that material was functioning extensively among Natufian communities.

For a long time, it was thought that the late Epipalaeolithic Natufian communities exploited the rich environment of the Mediterranean woodland zone of the Levant. Over recent decades several archaeologists have explored the semi-arid interior of Jordan, including the harsh basalt landscape of the Black Desert. There are locations within that landscape where there were springs or areas of inland drainage that produced seasonal wetlands. Shubayqa 1 is a recently excavated site on the edge of a mudflat, the location of a seasonal lake. The site was occupied throughout the late Epipalaeolithic period. They exploited mainly club-rush tubers; less than 10% of the plant remains were cereals or legumes, the founder-crops of the Neolithic (Arranz-Otaegui et al. 2018). Nevertheless, the inhabitants were producing flour both from the club-rush tubers and the small amounts of cereals, and baking bread.

Gazelle were hunted throughout the occupation and there was also an intensive exploitation of the plentiful migrating waterfowl that would have been present when the mudflat was flooded in the winter (Yeomans & Richter 2018). Small mammals such as hare were also hunted at Shubayqa, It seems likely that this was not an adaptation that was enforced by pressure on resources, but rather because people at Shubayqa were using a new technology, hunting with dogs (Yeomans et al. 2019). The intensive and efficient exploitation of a local environment in order to reduce mobility was a common feature in the Natufian, and Shubayqa, in the middle of the Black Desert, is no exception.

Beyond the South Levant

The Mediterranean woodland zone, which was so important for the Natufian in the south, extended through the north Levant, with similarly rich environmental resources. A number of sites that can be dated to the late Epipalaeolithic on account of the resemblance of their chipped stone assemblages to the south Levantine Natufian have been located the length of west Syria, but few have been excavated (Ibáñez et al. 2017). Most are situated in the foothills of the Anti-Lebanon mountain range that separates Syria from Lebanon. From surface survey, some sites show evidence of circular stone-walled structures, ground stone equipment for grinding and pounding, or, on exposed rock surfaces, shallow bedrock mortars. At one site in the far south of Syria, Qarassa, 12 circular structures, defined by the stones at the base of their walls, was set on the basalt bedrock in an arc around the edge a natural bowl.

On a headland overlooking the narrow floodplain of the Euphrates, about 80 km east of Aleppo, Abu Hureyra was excavated as part of a salvage archaeology programme in the early 1970s, when a major dam was under construction. It was planned to produce maximal samples of botanical and zoological data by a directorial trio (Moore et al. 1975, 2000). The prehistoric remains consisted of two phases with a period of abandonment between them. Abu Hureyra 1 is well dated by radiocarbon dating to the late Epipalaeolithic. It was a small settlement that was established shortly before the beginning of the Younger Dryas phase and it seems to have been abandoned shortly before the end of that phase. By contrast Abu Hureyra 2 was a large settlement dating to the Pre-Pottery Neolithic, and we shall return to it in a following chapter.

The excavators of Abu Hureyra used their brief opportunity to best advantage, undertaking a huge amount of intensive sieving and flotation of the deposits. The archaeological deposit representing the late Epipalaeolithic occupation was about 1 m thick. The earliest structures were small, circular huts whose floors were dug down into the natural subsoil. Later structures were built on the ground surface. The cultural material is clearly recognisable as practically identical to the late or final Natufian of the south Levant, and it was clearly a settlement established where its inhabitants could exploit the resources of several, very different and complementary ecological zones – the river for its large catfish, the narrow floodplain of the Euphrates, and the semi-arid landscape that was at the margins of the Mediterranean woodland zone, which would have been where the wild cereals, legumes and other plant foods would have been found. The analysis and interpretation of the largest collection of carbonised plant remains from any site in Southwest Asia occupied the archaeo-botanist Gordon Hillman and generations of his PhD students for decades. Hillman came to the conclusion on the balance of probabilities that the inhabitants had responded to the impact of the environmental deterioration of the Younger Dryas by beginning to cultivate cereals, in particular rye (Hillman et al. 2001; Moore et al. 2000; Moore & Hillman 1992). In their review of the published data, however, Colledge and Conolly have argued persuasively that Hillman's interpretation is dubious at several points, and that a more plausible interpretation is that the small-seeded plants that Hillman believed were 'weeds of cultivation' were in fact plants that had been intensively collected for their seeds to supplement the declining availability of the wild cereals (Colledge & Conolly 2010).

Just 40 km upstream there was another settlement, Tell Mureybet was established at almost exactly the time that Abu Hureyra was abandoned, that is, shortly before the end of the Younger Dryas phase. Having been first occupied around 10,200 BCE, the settlement of Tell Mureybet continued across the end of the Younger Dryas and for many centuries in the Pre-Pottery Neolithic. It therefore appears briefly in this chapter, and figures more in the following chapter. Like Abu Hureyra, the site was investigated in the early 1970s as a brief salvage excavation when the Tabqa dam. As with Abu Hureyra, the abundant material recovered in just two seasons of excavation occupied years of analysis and research before the full results could be brought together and integrated (Cauvin 1977; Ibañez 2008). Although Tell Mureybet occupied a virtually identical niche to Abu Hureyra, somehow the small community who settled there in the heart of the Younger Dryas survived and continued for more than 2000 years. In particular, there is no sign in their subsistence economy of difficulties because of a harsh Younger Dryas climatic impact on the available resources. The inhabitants gathered wild barley and rye, and also wild pistachio and plants native to the floodplain such as club-rush. The distinctive presence of barley and rye posed a question for the archaeo-botanist on the team, because these cereals would not have been part of the local ecology at that time. There were two options: either the inhabitants were harvesting the wild cereals where they were available, at least 60 km north of Tell Mureybet, or they were cultivating these wild species in an environment where they were not native. Particularly in the light of the charred cereals that he was studying from other sites of slightly later date (which we shall be meeting in the next chapter) the archaeo-botanist George Willcox reasoned that people had begun to cultivate the wild cereals in plots around their settlements (Willcox et al. 2008; Willcox & Savard 2011).

The last site in north Syria at which we should pause is the cave site of Dederi-yeh, situated in the far northwest corner of Syria, within the moist Mediterranean woodland environment of limestone hill country. The site has a long stratigraphy through the Middle Palaeolithic, within which important Neanderthal burials were found, but there is also a Natufian occupation in the mouth of the cave (Akazawa & Nishiaki 2017). At least five circular structures have been identified, one of which was destroyed by fire; in addition to the charred timbers of the building, valuable charred plant remains were found on the floor (Tanno et al. 2013), which were referred to in Chapter 2. This paper was a preliminary report on continuing research. There were wild wheats, a little wild barley, lentils, pea and bitter vetch, Pistacia nuts and almonds. The chipped stone assemblage is effectively Natufian and can be dated early in the late Epipalaeolithic. The presence of relatively uncommon bone tools, and of ground stone equipment and dentalium shell beads suggest that this was a base-camp or small settlement as we have seen in the southern Levant.

And beyond the Levant

Across the northern arc of the hilly flanks zone, through southeast Turkey and north Iraq there are no known Epipalaeolithic sites, except for two or three settlements, which, like Tell Mureybet, were first occupied during the Younger Dryas phase at the very end of the late Epipalaeolithic; and, like Tell Mureybet, they continued in occupation for many centuries in the early Pre-Pottery Neolithic. We shall meet those Tigris valley settlements in the next chapter.

Looking further east into the mountains of Iraqi Kurdistan, we will again encounter the intrepid Dorothy Garrod, whom we met in connection with her excavations in the caves of the Mount Carmel hills behind the Mediterranean coast. In the 1920s she investigated the stratified Palaeolithic in caves in the Zagros mountains (Garrod 1930). She found strata that she could recognise as Upper Palaeolithic, and also a microlithic industry that parallels the Epipalaeolithic industries of the Levant. The same Zarzian (it was named after the cave site Zarzi where Garrod had identified it) has been recognised in numerous sites In the mountainous northeast of Iraq, where the Taurus mountain range and its piedmont bend round to become the Zagros mountains, and in the Zagros valleys of western Iran. The radiocarbon dates for sites with Zarzian chipped stone assemblages begin around 14,000 BCE (after the Last Glacial Maximum, and close to the end of the early Epipalaeolithic in our terms), and they continue through the late Epipalaeolithic. Those who have studied the Upper Palaeolithic and Zarzian industries of the region at first hand have remarked that the Zarzian is a development from the Upper Palaeolithic tradition; but there is a sizeable gap in the radiocarbon dates, coinciding with the Last Glacial Maximum, which would probably have made the intermontane valleys of the Zagros uninhabitable.

There is a pair of sites in a valley in the Zagros in northeast Iraq that illustrates the potential of the Zarzian communities to emulate their late Epipalaeolithic Natufian contemporaries in the Levant. Ralph Solecki's excavations at Shanidar cave in the 1950s produced a stratum, which, in the very early days of radiocarbon dating, was only rather imprecisely dated late in the final Epipalaeolithic (Solecki 1963; Solecki & Solecki 1983). The succeeding level was called 'Proto-Neolithic' by Solecki and it was compared to Zawi Chemi, a small open site nearby for which he obtained a radiocarbon date at the very end of the Epipalaeolithic period (Solecki 1981). Whereas the earlier level had no ground stone implements, the upper level produced cobbled floors, some ground stone implements and a cemetery of burials (Solecki et al. 2004). Remains of some 35 individuals, ranging from infants, through children and adolescents to adults, both male and female, were recovered from 26 graves. Many of the bodies had been folded tightly into small pits. Other bodies were represented by only a skull or a few bones, and were secondary burials; in some cases, parts of two or more different individuals were together in the same grave. A number of the bodies were accompanied by one or two stone beads; in one case the bead was a made from a green copper mineral.

The nearby open site of Zawi Chemi reinforces and extends the information from the cave site. In the last millennium or so of the Epipalaeolithic period the Zarzian changed; from the brief but informative report, we can see a trend towards sedentism, the construction of permanent buildings, and the use of personal ornament that echoes what we have seen in the Natufian. The site was below the cave site, on the valley floor at a lower altitude (425 m above sea level), and it took the form of a low mound of occupation debris. The stratification was up to 3 m thick, a powerful indication of long-term and continuous occupation and the use of a good deal of building material. The architecture was eroded and almost no complete buildings were found, but there were large pits, arcs of stone-built wall foundations, and evidence of several episodes of rebuilding or remodelling. Near the top of the stratigraphy, one circular stone construction was recovered.

The excavators also noted the presence of beads and pendants made of steatite, bone and animal teeth. More than 200 ground stone implements were recorded, but, unfortunately, the excavations in the Shanidar valley in the 1950s were too early for the use of flotation and wet sieving which might have produced information on the plant foods that were being collected, harvested and processed. Animal bone was recovered, however, and through most of the stratigraphy deer, followed by wild goat and sheep, dominate the faunal remains. In the latest level, the balance of species changed: deer became less common and wild sheep became the dominant species, followed by goat. As we saw in Chapter 2, Dexter Perkins Jr. (1964), who studied the animal bones, thought that the shift in the relative importance of species and the large numbers of immature sheep indicated that domestication was under way. Renewed study by Melinda Zeder has identified an over-representation of immature young male sheep; and, as Perkins had noted, that was different from the spectrum for goats in the earlier levels (Zeder 2008, 2011). Zeder explains that what we see at Zawi Chemi may be a deliberate strategy of focusing on the hunting of male rather than female sheep in order to minimise the effect of hunting on the breeding of the wild stock.

Flannery had the basis for his formulation of his broad-spectrum revolution theory (Flannery 1969), constructed on the framework of a paper by Lewis Binford (1968). Binford had proposed that, where hunter-gatherers lived in 'optimal zones' (that is, where rich resources encouraged population growth, as in the Zagros valleys and around the hilly flanks of the Fertile Crescent), population would have to expand outwards into less optimal environmental conditions. Binford's argument suggested that manipulation of wild resources would be necessary in the marginal zone beyond the optimal conditions. For Flannery, the overspill of hunter-gatherer population from the Khorramabad valley moved out onto the alluvial plain, where it was necessary for them to find ways to compensate for the poor availability of wild resources, hence the cultivation of wild cereals and herding of wild sheep and goats that they had brought with them.

Having wheeled around the arc of the hilly flanks zone, we should briefly turn our attention westwards towards Anatolia. Until recently the only late Pleistocene sites were in the south of Turkey, in caves overlooking the coastal plain near Antalya (Kara'in and Öküzini). This plain behind Antalya is geographically cut off from the high plateau of central Anatolia by the Taurus mountains, and the climatic conditions of the Mediterranean coast are quite different from those of the plateau. There are deposits that date from the Middle, Upper and Epipalaeolithic periods, as well as the aceramic Neolithic and later. Öküzini has a succession of deposits that date through the Upper Palaeolithic and Epipalaeolithic into the Holocene (Otte et al. 1995). As elsewhere in the hilly flanks zone, the chipped stone industry turns to the production of microblades and microliths in the Epipalaeolithic, and the late Epipalaeolithic even includes some lunates, the characteristic microlith of the late Epipalaeolithic Natufian in the Levant. Among the chipped stone, there are pieces of obsidian from central Anatolia, showing that the Epipalaeolithic hunter-gatherers were not living in isolation from the wider cultural world.

There is one remarkable site in central Anatolia that is firmly dated to the late Epipalaeolithic period. Pınarbaşı is about 30 km southeast of Çatalhöyük on the southern edge of the Konya plain, and I have a personal connection with the site: it was the last excavation where I worked in the field (Baird 2012; Baird et al. 2013). In fact, Pınarbaşı consists of a cluster of sites of different periods at the tip of a line of low limestone hills that stretch out from the base of the black volcanic massif of Karadağ into the Konya plain. As its Turkish name indicates, there was a spring of fresh water there, and beyond the spring and its pool there was an extensive shallow seasonal lake and reedbeds. (I use the past tense, because the spring and the seasonal flooding of the reedbeds has dried up in recent decades).

In one of the several rock-shelter sites at the foot of the cliffs that overlook the spring there was a small Pre-Pottery Neolithic occupation. Beneath a deep layer of almost sterile material that was probably fragments of frost-shattered rock from the cliff above, the excavation came to a rich, dark, loamy deposit with charcoal traces of hearths, plentiful animal bone, and a microlithic chipped stone industry whose most striking characteristic was the copious numbers of crescentic microliths immediately reminiscent of the Natufian lunates of the southern Levant. Radiocarbon dates subsequently confirmed that this occupation of the rock-shelter dated to the late Epipalaeolithic. It was easy to appreciate why the location would have appealed to Epipalaeolithic hunter-gatherers. The rock-shelter looked out over an extensive pool or small lake that (within living memory) offered large carp; beyond that, almost to the horizon was seasonal shallow water and reedbeds. In addition to the regular wild-life, Pınarbaşı was a stopping point that attracted many migrant birds. Beyond the reedbeds the Konya plain offered opportunities for hunting large game, especially aurochs, wild cattle. And the limestone hills and the nearby basalt volcanic massif presented more varied ecological zones for hunting, trapping and gathering. It is a classic location of the kind that attracted Epipalaeolithic hunter-gatherer groups to spend longer seasonal stays or even year-round occupations.

Beneath some depth of accumulated occupation deposit, two burials were found. The first of them was an adult male who was buried lying on his back. The grave had been revisited and the cranium and jawbone had been removed, leaving behind a few loose teeth. There are rare examples of this practice of retrieving the skull from a burial in the Natufian late Epipalaeolithic in the Levant. Although it seems that the rock-shelter was a seasonal base-camp rather than being permanently occupied, clearly the precise location of the grave and the head within the grave were remembered, or perhaps there had been some kind of a marker. The second grave dates some time – perhaps decades, perhaps centuries – after the first burial (and part of a third burial lay within the small trench). The intact skeleton was that of a mature adult male, accompanied by a remarkable cache of grave goods, placed behind the head and covered with red ochre. The body had been wrapped in a reed mat, identified by surviving phytoliths. The cache of grave goods consisted of four Nassarius marine shell beads, three bone beads and 140 Dentalium marine shells, all covered with red ochre and placed within an upturned tortoise shell. Despite our best efforts with flotation and wet sieving, there were no surviving plant remains. The stable isotope analyses of the two human skeletons produced very unusual results, which were interpreted as indicating a diet that was mainly composed of meat from the aurochs that roamed the plain, fish from the immediately adjacent lake, and birds (Baird et al. 2013: 197–200).

Our last stop in our tour of the late Epipalaeolithic is on the island of Cyprus. The site of Akrotiri Aetokremnos was found on the British airforce base area by the young son of a serving officer, who drew it to the attention of the amateur archaeological group on the base area, who in turn brought in Alan Simmons, a US Neolithic specialist archaeologist, to excavate the site (Simmons 1991, 1999, 2012). Aetokremnos is a collapsed rock-shelter in a steep slope overlooking the sea on the south coast of Cyprus. The excavations produced two remarkable and wholly unexpected results: there were clear traces of human occupation in the shape of hearths and chipped stone tools, but there were also thousands of bones of the extinct Cypriot pygmy hippopotamus; and the radiocarbon dates placed the human occupation in the last millennium of the late Epipalaeolithic period. Up till that time, the earliest accepted archaeological evidence of humans in Cyprus belonged to the island's Pre-Pottery Neolithic Khirokitia settlement, contemporary with the Pottery Neolithic on the Levantine mainland. The dates from Aetokremnos caused a stir: many archaeologists found it difficult to accept that the first colonists had sailed to Cyprus before the end of the Palaeolithic.

So much for the date of the occupation at Aetokremnos: the pygmy hippopotamus bones proved even more contentious. Alan Simmons' account of the site proposed that the mass of pygmy hippopotamus bones in a rock-shelter halfway down a cliff on the coast of Cyprus were the by-product of butchering by the Epipalaeolithic group who had lived there. Zoologists who knew of the Pleistocene fauna of Cyprus generally date the extinction of the island's pygmy hippopotamus (and elephant) considerably before the human occupation of Aetokremnos; and several archaeologists have questioned Simmons' interpretation, arguing that the pygmy hippopotamus bones formed a separate, earlier stratum than the use of the rock-shelter by a group of hunter-gatherers. The subject remains contentious, but while studying the animal bones that were firmly associated with the hearths and the chipped stone tools, Jean-Denis Vigue identified several bones of wild pig, a species not native to the island. Vigne found that the pig bones were significantly smaller than contemporary wild pig on the mainland. Their reduced size, he has argued, implies that their ancestors had been brought to the island some time earlier and had been managed under human control for more than a thousand years. It is not so surprising that wild pigs were under human management at the close of the late Epipalaeolithic period. Those earlier hunter-gatherers had deliberately manipulated the island environment so that it could support their way of life. Later, in the early Pre-Pottery Neolithic, we shall see that later arrivals introduced other species, some to be managed and herded and others to be available for hunting. Following from Vigne's interpretation, it follows that at some time in the Epipalaeolithic period hunter-gatherers deliberately stocked the island with wild pig. There has been a series of unexpected discoveries of earlier and earlier Neolithic sites in Cyprus, and we now need even earlier, Epipalaeolithic period, sites if Vigne's hypothesis is to be validated.

There are now late Epipalaeolithic sites known from around the hilly flanks zone, as well as in central southern Anatolia; however, the southern Levant still dominates the picture because of the concentration of research over so many years. The idea that the milder climate and generous resources of the Mediterranean woodland zone of the Levant encouraged Epipalaeolithic hunter-gatherers to lay the groundwork for the succeeding Pre-Pottery Neolithic period and the development of farming no longer holds good. We have seen that there were communities all around the region that were developing ways to concentrate their lives around seasonal resources, especially storable resources, living together in larger numbers, creating settlements of permanent buildings, and reducing their reliance on mobility. In the past we have been encouraged to think that the harvesting and storage of wild cereals that has been identified at Epipalaeolithic sites in the Levant was the preliminary stage leading to their cultivation and domestication in the Neolithic; but we now know that late Epipalaeolithic communities beyond the Levant were living in a similar way but were not reliant to any extent on wild cereals. Since the recognition of the Younger Dryas phase at the end of the Pleistocene period, researchers have been looking for traces of its impact on the landscape and resources of the final Epipalaeolithic peoples of southwest Asia. At first it was thought that the Younger Dryas impact explained the apparent decline of Natufian settlements like Eynan in Israel; but, as we have seen here, we now know that the Natufian communities continued to thrive, even if they stopped building massive semi-subterranean buildings. And we are beginning to get reports of salvage archaeology excavations at a string of remarkable early Pre-Pottery Neolithic settlements along the Tigris valley in southeast Anatolia that were in fact founded several centuries earlier, in the Younger Dryas phase.

References

Akazawa, T., & Nishiaki, Y. (2017). The Palaeolithic Cultural Sequence of Dederiyeh Cave, Syria. In O. Bar-Yosef & Y. Enzel (Eds.), *Quaternary of the Levant: Environments, Climate Change, and Humans* (pp. 307–14). Cambridge: Cambridge University Press.

Arranz-Otaegui, A., González-Carretero, L., Roe, J., & Richter, T. (2018). "Founder Crops" v. Wild Plants: Assessing the Plant-Based Diet of the Last Hunter-Gatherers in Southwest Asia. *Quaternary Science Reviews*, 186, 263–83.

Baird, D. (2012). Pınarbaşı: From Epipalaeolithic Camp-Site to Sedentarising Village in Central Anatolia. In M. Özdoğan, N. Başgelen, & P. Kuniholm (Eds.), *The Neolithic in Turkey. New Excavations and New Research: Central Turkey* (pp. 181–218). Istanbul: Arkeoloji ve Sanat Yayinlari.

Baird, D., Asouti, E., Astruc, L., Baysal, A., Baysal, E., Carruthers, D., . . . Pirie, A. (2013). Juniper Smoke, Skulls and Wolves' Tails. The Epipalaeolithic of the Anatolian Plateau in Its South-West Asian Context; Insights from Pınarbaşı. *Levant*, 45(2), 175–209. doi:10.1 179/0075891413Z.00000000024

Bar-Yosef, O., & Valla, F. (1991). *The Natufian Culture in the Levant*. Ann Arbor, MI: International Monographs in Prehistory, Archaeological Series Volume 1.

Bar-Yosef, O., & Valla, F. R. (2013). *Natufian Foragers in the Levant: Terminal Pleistocene Social Changes in Western Asia*. Ann Arbor, MI: International Monographs in Prehistory.

Belfer-Cohen, A., & Goring-Morris, A. N. (2013). Breaking the Mould: Phases and Facies in the Natufian of the Mediterranean Zone. In O. Bar-Yosef & F. R. Valla (Eds.), *Natufian Foragers in the Levant: Terminal Pleistocene Social Changes in Western Asia* (pp. 544–61). Ann Arbor, MI: International Monographs in Prehistory.

Binford, L. R. (1968). Post-Pleistocene adaptations. In L. R. Binford & S. R. Binford (Eds.), *New Perspectives in Archaeology* (pp. 313–341). Chicago: University of Chicago Press.

Bouchud, J. (1987). *La faune de Mallaha (Eynan), Israël*. Paris: Mémoires et Travaux du CRFJ No. 4. Association Paléorient.

Boyd, B. (1995). Houses and Hearths, Pits and Burials: Natufian Mortuary Practices at Mallaha (Eynan), Upper Jordan Valley. In S. Campbell & A. Green (Eds.), *The Archaeology of Death in the Ancient Near East* (pp. 17–23). Oxford: Oxbow Books.

Cauvin, J. (1977). Les fouilles de Mureybet (1971–1974). et leur signification pour les origines de la sédentarisation au Proche-Orient. *Annual of the American Schools of Oriental Research*, 44, 19–48.

Colledge, S., & Conolly, J. (2010). Reassessing the evidence for the cultivation of wild crops during the Younger Dryas at Tell Abu Hureyra, Syria. *Environmental Archaeology*, 15(2), 124–38.

Dubreuil, L. (2004). Long-Term Trends in Natufian Subsistence: A Use-Wear Analysis of Ground Stone Tools. *Journal of Archaeological Science*, 31(11), 1613–29. doi:10.1016/j.jas.2004.04.003|ISSN 0305-4403

Flannery, K. V. (1969). The origins and ecological effects of early domestication in Iran and the Near East. In P. J. Ucko & G. W. Dimbleby (Eds.), *The Domestication and Exploitation of Plants and Animals* (pp. 73–100). London: Duckworth.

Garrod, D. A. E. (1930). The Palaeolithic of Southern Kurdistan: Excavations in the Caves of Zarzi and Hazar Merd. *Bulletin of the American School of Prehistoric Research*, 6, 8–43.

Garrod, D. A. E., Bate, D. M. A., McCown, T. D., & Keith, A. (1937). *The Stone Age of Mount Carmel. Joint Expedition of the British School of Archaeology in Jerusalem and the American School of Prehistoric Research (1929–1934)* (Vol. 2 vols.). Oxford: The Clarendon Press.

Grosman, L., & Belfer-Cohen, A. (2022). Moving On: Natufian After-Life Phases and Stages. Hilazon Tachtit Cave as a Case Study. In D. Ackerfeld & A. Gopher (Eds.), *Dealing with the Dead - Studies on Burial Practices in the Pre-Pottery Neolithic Levant*. Berlin: Ex Oriente.

Grosman, L., & Munro, N. D. (2016). A Natufian Ritual Event. *Current Anthropology*, 57(3), 311–31. doi:10.1086/686563

Grosman, L., Munro, N. D., & Belfer-Cohen, A. (2008). A 12,000-Year-Old Shaman Burial from the Southern Levant (Israel). *Proceedings of the National Academy of Sciences of the United States of America*, 105, 17665–9. doi:10.1073/pnas.0806030105

Henry, D. O. (1989). *From Foraging to Agriculture: The Levant at the End of the Ice Age*. Philadelphia: University of Pennsylvania Press.

Henry, D. O. (1995). *Prehistoric Cultural Ecology and Evolution: Insights from Southern Jordan*. New York, NY; London: Plenum Press.

Henry, D. O. (2013). The Natufian and the Younger Dryas. In O. Bar-Yosef & F. R. Valla (Eds.), *Natufian Foragers in the Levant: Terminal Pleistocene Social Changes in Western Asia* (pp. 584–610). Ann Arbor, MI: International Monographs in Prehistory.

Hillman, G. C., Hedges, R., Moore, A. M. T., Colledge, S., & Pettitt, P. (2001). New Evidence of Lateglacial Cereal Cultivation at Abu Hureyra on the Euphrates. *The Holocene*, 11(4), 383–93.

Ibañez, J. J. (2008). *Le site néolithique de Tell Mureybet: en hommage à Jacques Cauvin (2 vols)* (Vol. British Archaeological Reports, International Series1843). Oxford: Archaeopress.

Ibáñez, J. J., González-Urquijo, J., & Terradas, X. (2017). The Natufian Period in Syria. In Y. Enzel & O. Bar-Yosef (Eds.), *Quaternary of the Levant: Environments, Climate Change, and Humans* (pp. 709–14). Cambridge: Cambridge University Press.

Moore, A. M. T., Hillman, G., & Legge, A. (1975). The Excavation of Tell Abu Hureyra in Syria: A Preliminary Report. *Proceedings of the Prehistoric Society*, 41, 50–77.

Moore, A. M. T., & Hillman, G. C. (1992). The Pleistocene to Holocene Transition and Human Economy in Southwest Asia: The Impact of the Younger Dryas. *American Antiquity*, 57, 482–94.

Moore, A. M. T., Hillman, G. C., & Legge, A. J. (2000). *Village on the Euphrates: From Foraging to Farming at Abu Hureyra*. Oxford: Oxford University Press.

Munro, N. D., & Grosman, L. (2010). Early Evidence (ca. 12,000 B.P.) for Feasting at a Burial Cave in Israel. *Proceedings of the National Academy of Sciences of the United States of America*, 107(35), 15362–6.

Otte, M., Yalcinkaya, I., Leotard, J. M., Kartal, M., Bar-Yosef, O., Kozłowski, J., . . . Marshack, A. (1995). The Epi-Palaeolithic of Öküzini Cave (SW Anatolia) and Its Mobiliary Art. *Antiquity*, 69(266), 931–44.

Perkins, D. (1964). Prehistoric Fauna from Shanidar, Iraq. *Science*, 144, 1565–6.

Perrot, J., Ladiray, D., & Solivères-Massei, O. (1988). *Les Hommes de Mallaha (Eynan) Israël*. Paris: Association Paléorient.

Samuelian, N. (2013). A Study of Two Natufian Residential Complexes: Structures 200 and 203 at Eynan (Ain Mallaha), Israel. In O. Bar-Yosef & F. o. R. Valla (Eds.), *Natufian Foragers in the Levant: Terminal Pleistocene Social Changes in Western Asia* (pp. 172–84). Ann Arbor, MI: International Monographs in Prehistory.

Simmons, A. (1991). Humans, Island Colonization and Pleistocene Extinctions in the Mediterranean: The View from Akrotiri Aetokremnos, Cyprus. *Antiquity*, 65, 857–69.

Simmons, A. (1999). *Faunal Extinction in an Island Society: Pygmy Hippopotamus Hunters of Cyprus*. New York, NY: Kluwer Academic/Plenum.

Simmons, A. (2012). Mediterranean Island Voyages. *Science*, 338(6109), 895–7. doi:10.1126/science.1228880

Solecki, R. L. (1981). *An Early Village Site at Zawi Chemi Shanidar*. Malibu, CA: Biblioteca Mesopotamica, Vol. 13. Undena Publications.

Solecki, R. L., & Solecki, R. S. (1983). Late Pleistocene - Early Holocene Cultural Traditions in the Zagros and the Levant. In T. C. Young, P. E. L. Smith, & P. Mortensen (Eds.), *The Hilly Flanks and beyond: Essays on the Prehistory of Southwestern Asia* (Vol. Studies in ancient oriental civilization, No. 36, pp. 123–37). Chicago, IL: Oriental Institute of the University of Chicago.

Solecki, R. S. (1963). Prehistory in the Shanidar Valley, Northern Iraq. *Science*, 139(3551), 179.

Solecki, R. S., Solecki, R. L., & Agelarakis, A. G. (2004). *The Proto-Neolithic Cemetery in Shanidar Cave*. College Station: Texas.

Tanno, K., Willcox, G., Muheisen, S., Nishiaki, Y., Kanjo, Y., & Akazawa, T. (2013). Preliminary Results from the Analyses of Charred Plant Remains from a Burnt Natufian Building at Dederiyeh Cave in Northwest Syria. In O. Bar-Yosef & F. o. R. Valla (Eds.), *Natufian Foragers in the Levant: Terminal Pleistocene Social Changes in Western Asia* (pp. 83–7). Ann Arbor, MI: International Monographs in Prehistory.

Tchernov, E., & Valla, F. (1997). Two New Dogs, and Other Natufian Dogs, from the Southern Levant. *Journal of Archaeological Science*, 24(1), 65–95.

Valla, F. (1984). *Les industries de silex de Mallaha (Eynan) et du Natoufien dans le Levant*. Paris: Association Paléorient.

Valla, F. (1988). Aspects du sol de l'abri 131 de Mallaha (Eynan), Israel. *Paléorient*, 14(2), 283–96.

Valla, F., Khalaily, H., Samuelian, N., Bocquentin, F., Bridault, A., & Rabinovich, R. (2017). Eynan (Ain Mallaha). In O. Bar-Yosef & Y. Enzel (Eds.), *Quaternary of the Levant: Environments, Climate Change, and Humans* (pp. 295–302). Cambridge: Cambridge University Press.

Valla, F., Khalaily, H., Samuelian, N., Bridault, A., Rabinovich, R., Simmons, T., . . . Ash-kenazi, S. (2013). The Final Natufian Structure 215–28 at Mallaha (Eynan), Israel: An Attempt at Spatial Analysis. In O. Bar-Yosef & F. o. R. Valla (Eds.), *Natufian Foragers in the Levant: Terminal Pleistocene Social Changes in Western Asia* (pp. 146–71). Ann Arbor, MI: International Monographs in Prehistory.

Weinstein-Evron, M., Kaufman, D., Bachrach, N., Bar-Oz, G., Bar-Yosef Mayer, D., Cham, S., . . . Weissbrod, L. (2007). After 70 Years: New Excavations at the el Wad Terrace, Mount Carmel, Israel. *Jurnal of the Israel Prehistoric Society*, 37, 37–134.

Weinstein-Evron, M., Kaufman, D., & Yeshurun, R. (2013). Spatial Organization of Natufian el Wad through Time: Combining the Results of Past and Presemt Excavations. In O. Bar-Yosef & F. o. R. Valla (Eds.), *Natufian Foragers in the Levant: Terminal Pleistocene Social Changes in Western Asia* (pp. 88–106). Ann Arbor, MI: International Monographs in Prehistory.

Weissbrod, L., Marshall, F. B., Valla, F. R., Khalaily, H., Bar-Oz, G., Auffray, J.-C., . . . Cucchi, T. (2017). Origins of House Mice in Ecological Niches Created by Settled Hunter-Gatherers in the Levant 15,000 y Ago. *Proceedings of the National Academy of Sciences*, 114(16), 4099–104. doi:10.1073/pnas.1619137114

Willcox, G., Fornite, S., & Herveux, L. (2008). Early Holocene Cultivation before Domestication in Northern Syria. *Vegetation History and Archaeobotany*, 17(3), 313–25.

Willcox, G., & Savard, M. (2011). Botanical Evidence for the Adoption of Cultivation in Southeast Turkey. In M. Özdogan, N. Başgelen, & P. Kuniholm (Eds.), *The Neolithic in Turkey, New Excavations and New Research* (pp. 267–80). Istanbul: Archaeology and Art Publications.

Yeomans, L., Martin, L., & Richter, T. (2019). Close Companions: Early Evidence for Dogs in Northeast Jordan and the Potential Impact of New Hunting Methods. *Journal of Anthropological Archaeology*, 53, 161–73. doi:10.1016/j.jaa.2018.12.005

Yeomans, L., & Richter, T. (2018). Exploitation of a Seasonally Abundant Resource: Bird Hunting during the Late Natufian at Shubayqa 1. *International Journal of Osteoarchaeology*, 28(2), 95–108.

Yeshurun, R., Bar-Oz, G., & Weinstein-Evron, M. (2014). Intensification and Sedentism in the Terminal Pleistocene Natufian Sequence of el Wad Terrace (Israel). *Journal of Human Evolution*, 70, 16–35. doi:10.1016/j.jhevol.2014.02.011

Zeder, M. A. (2008). Domestication and Early Agriculture in the Mediterranean Basin: Origins, Diffusion, and Impact. *Proceedings of the National Academy of Sciences of the United States of America*, 105(33), 11597–604. doi:10.1073/pnas.0801317105

Zeder, M. A. (2011). The Origins of Agriculture in the Near East. *Current Anthropology*, 52(S4), S221–35. Retrieved from http://www.jstor.org/stable/10.1086/659307

6 Early Pre-Pottery Neolithic – Transforming Their World

At this point in our chronological survey we leave the Pleistocene period and enter the Holocene: in archaeological terminology, we leave the Palaeolithic and enter the Neolithic; but these apparently big steps are the consequence of our inherited terminology. We shall see that there was a very obvious step-change in the archaeology between the late Epipalaeolithic and the early Pre-Pottery Neolithic, which is something that we shall have to evaluate. There was a climate change, as the brief, final cooler and drier phase of the Pleistocene, the Younger Dryas, rapidly gave way to a Holocene optimum. Temperatures recovered and became generally as they are today, and rainfall was, in some regions at least, a little more than at present. How these climatic changes impacted on the varied environments of the region and how they were reflected in the archaeology is still debated and remains unclear.

In the Levant, there was a rapid (and for archaeologists very convenient) transition in the chipped stone industries: the neat and distinctive bladelets and microliths that had typified the assemblages of the whole of the Epipalaeolithic period were replaced by equally finely made blades and distinctive forms of projectile point. Beyond the Levant, the transition from the Epipalaeolithic into the Neolithic is not clearly recognisable in the chipped stone industries, and can only defined by radiocarbon dating, either before or after about 9600 BCE. And in other aspects of material culture, almost everything that is found on early Pre-Pottery Neolithic sites, whether buildings, burials, grinding stones or mortars and pestles, can be found on a late Epipalaeolithic site somewhere. Nevertheless, the early Pre-Pottery Neolithic was, as its technical label suggests, new. Few late Epipalaeolithic sites continued into the early Pre-Pottery Neolithic, and conversely almost all early Pre-Pottery Neolithic sites are new foundations. Most sites of the Epipalaeolithic period are in caves and rock-shelters; open-air settlements are relatively rare. In the early Pre-Pottery Neolithic stratified, densely built settlements are the norm. Although early Pre-Pottery Neolithic settlements are mostly small in comparison with the settlements of the later Pre-Pottery Neolithic, they are significantly larger than the late Epipalaeolithic Natufian sites, and there is clearly a major investment in their architecture.

The discovery of – one could say the invention of – the Pre-Pottery Neolithic began with Kathleen Kenyon's excavations in the 1950s at the site of ancient Jericho,

DOI: 10.4324/9781351069281-7

Tell es-Sultan, in the Jordan valley close to the north end of the Dead Sea. Kenyon was by no means the first to excavate there, but she wanted to use modern excavation methods to resolve the question of whether the Biblical account of the destruction of the city by Joshua and the Israelites could be documented in the Bronze Age walled city remains. She also knew that there were metres of mysterious pre-Bronze Age remains, and she was determined to document the stratigraphy, and date it using the new techniques of radiocarbon dating (Kenyon 1957; Kenyon & Holland 1981). Her meticulous contextual recording and publication of the archaeological material and of the stratigraphy was backed up by a triple series of samples sent to no less than three leading radiocarbon dating laboratories. She showed that ancient Jericho has been occupied in the Neolithic period, which the radiocarbon dating showed was of much longer duration and was several thousand years earlier than had been cautiously estimated.

At the base of the stratigraphy there was a thin layer of poorly preserved structural remains from which she obtained chipped stone tools that were recognisably Natufian; but it was clear that there was a hiatus before the Neolithic settlement was founded. Kenyon noted a clear division of the deep Neolithic stratigraphy into three occupations. Only in the thin uppermost layers was there pottery, as was expected at that time of the Neolithic. By far the greater part of the deep stratigraphy seemed to be Neolithic, but without pottery, for which Kenyon invented the label Pre-Pottery Neolithic. That Pre-Pottery Neolithic was itself recognisably in two parts; there were differences in the chipped stone assemblages, in the shape of the mud-brick houses (circular in the early half, rectangular in the later half), and in the shape of the mud bricks with which they were built. So Kenyon called them PPNA and PPNB; and she and a soil scientist colleague at the Institute of Archaeology in London very cleverly identified a natural soil horizon that had developed, indicating a break in the occupation of the mound. While still engaged in the Jericho excavations, Kenyon wrote a general book on the archaeology of the 'Holy Land' (Kenyon 1960), in which she extended her understanding of the Jericho Neolithic stratigraphy to the whole region, defining a PPNA culture followed by a PPNB, labels that are still widely used for the earlier and later Pre-Pottery Neolithic periods (although I avoid their use because of their ambiguity of reference both to periods of time and to define prehistoric 'cultures').

Most excavations have concentrated on areas in the heart of the settlement, but Kenyon dug deep trenches that cut into the steep sloping sides of the Jericho mound. That enabled her to reach the deeply buried earliest strata of the site. At three places, in the north, west and south sides of the mound, she encountered a massive stone wall that seems to have surrounded the settlement. Outside the wall there was a rock-cut ditch, reinforcing the idea that these were the settlement's defences. The area enclosed was only very slightly less than the area within the Bronze Age city ramparts. The settlement was home to a large community, and they repeatedly engaged in maintaining their circuit wall. As repeated rebuilding and replacement of the mud – brick houses within the settlement had caused the mound to grow, the community had raised the height of their circuit wall. In the largest trench, on the west side of the mound, there was an even more imposing

Figure 6.1 Jericho, the early PPN stone tower against the inner side of the circuit wall. A man is standing at the doorway at the base of the tower, which leads to the steep internal staircase. Another man stands on the top of the tower, where the staircase emerges. (Kenyon & Holland 1981, Vol. 3, Pl. 9, with permission)

(and equally puzzling) construction, a massive circular tower of solid stone and mud mortar. The tower was attached to the inner side of the circuit wall, which means that it was not a conventional defensive bastion. Indeed, it is difficult to think of the circuit wall as a defensive necessity, since there are no other major settlements that could be potential aggressors (Figure 6.1).

The Neolithic levels at Jericho were deeply buried by later occupation, so Kenyon was able to uncover only limited areas of the settlement within its circuit wall, showing that the interior was packed with simple houses. Especially in the PPNB levels there were numerous human burials among the houses, many of which lacked skulls. The number of intramural burials was surprising, but it was no match for the discovery of several caches of human skulls. And some of those skulls were found to have complete faces modelled in clay, sometimes with traces of colour on the surfaces. The modelled skulls of Jericho rapidly became the icon, as it were, of the Pre-Pottery Neolithic. Jericho remains a unique settlement, which existed for several centuries in the early Pre-Pottery Neolithic, was reoccupied for several more centuries in the later Pre-Pottery Neolithic, and for a third time in the Pottery Neolithic period. It also showed that the settlements of the Pre-Pottery Neolithic were not simply villages, collections of small houses. The discovery of many intramural burials, the missing crania, the caches of skulls, and the modelling of faces on some of them was a startling introduction to the customs of the later Pre-Pottery Neolithic communities in the Levant. And, finally, the absence of farmed crops and herded animals in the earlier Pre-Pottery Neolithic levels showed very clearly that, in the Levant and southwest Asia at least, the simple identification of farming with the Neolithic did not apply. Kenyon's excavations and the analysis of the animal bones in particular indicated the route to new fields of study of domestication in both plants and animals.

To see inside an early Pre-Pottery Neolithic settlement in the Jordan valley we can turn to Netiv Hagdud, which was occupied for only a part of the early Pre-Pottery Neolithic period. Because there was no later occupation it was possible to expose about 500 m^2 of the approximately 15,000 m^2 site when it was excavated in the 1980s (Bar-Yosef et al. 1991). There were oval and sub-circular houses similar to those found at Jericho. Each was constructed in a shallow cavity that was excavated into earlier levels. The base of the walls was formed of large stones set on edge, above which they were constructed of unbaked, plano-convex mud bricks (flat on the under-surface, slightly curved on the upper side), exactly like those of early Pre-Pottery Neolithic A Jericho. The houses were quite small, only 4–9 m across, and only one of them had any internal division of the space. That house had been destroyed in a fire. On its floor there were the fragmentary remains of at least three human skulls, and there were pestles, broken mortars, stone bowls, and a number of polished pebbles, but only a very few of the normal flint tools. The excavators believed that the fire was not accidental, and that it was a building with a different function from the other houses. Altogether some 22 burials were found within the settlement. The bodies had been buried in tightly contracted positions in shallow pits. The bodies of young children were found intact, but the crania had been removed from all the adult bodies. The fact that the mandibles (lower jaw) were found with the rest of the skeleton indicates that the bodies had been buried intact, but the burials had been re-opened after some interval to retrieve the crania. This was another demonstration of customs concerning the treatment of the dead, which could be documented rarely in the late Epipalaeolithic, becoming more elaborate and more common through the early and then the later Pre-Pottery Neolithic.

By the time of the excavations at Netiv Hagdud in the 1980s, flotation together with wet and dry sieving had become established practice in order to recover carbonised plant remains and lots of the smallest bone, as well as the smallest artefacts. Hence we know that the inhabitants of Netiv Hagdud collected more than 50 species of plants, including quantities of wild barley and some emmer. The team concluded that the inhabitants had begun small-scale cultivation, of which we will hear more when we come to Jerf el Ahmar. The presence of a variety of heavy grinding and pounding tools signals the importance of the large and small-seeded plants in the diet. The range of fauna represented at the site is extraordinary, but typical of a broad-spectrum hunting, fishing and trapping strategy. There are migratory ducks and geese, quail, frogs, fish, tortoise, fox, hare, and many bird species that are not found in the region today, where there is no longer wetland and open water. The main food species hunted was the gazelle, plus plenty of wild boar, some fallow deer, a few ibex and the occasional hartebeest to supplement the meat diet.

Before we leave the south Levant we should take a look at WF16, a site with a code but no name, in the south of Jordan. The site was one of a number that were found in 1996 in a survey of Wadi Faynan, a steep, dry valley leading down westwards from the limestone highlands into the Wadi Araba, that part of the great rift valley between the Dead Sea and the cities of Eilat and Aqaba at the head of the Red Sea. Preliminary soundings established that this was a small settlement that was occupied exclusively within the early Pre-Pottery Neolithic period, providing the opportunity for excavating an extensive exposure of most of the settlement. Open area excavations took place between 2008 and 2010 (Finlayson et al. 2011; Mithen et al. 2018) (Figure 6.2).

WF16 was a dense cluster of semi-subterranean oval structures whose walls were formed of pisé or adobe, that is a mixture of mud and plant material, supporting timber frames for flat roofs. The settlement was founded as the late Epipalaeolithic was transitioning into the earliest Pre-Pottery Neolithic period, around 10,000 BCE. It was occupied for almost 2000 years, with a peak of activity around 9200 BCE. The structures varied in size; some were houses, others were places for storage, and there were also buildings that were identified as workshops for making beads. There was also one especially large structure that seems to have been built to accommodate social gatherings and performances. (There are now examples of large, circular communal buildings at early Pre-Pottery Neolithic settlements around the arc of the hilly flanks zone as far as Tell Asiab in western Iran, as well as in Cyprus.) This was a truly huge building, around 19 m in diameter. Its mud-plastered floor was well below the surrounding ground level. Around its perimeter there was a sort of 'bench' that was more than a metre broad and half a metre tall. The floor was bisected by a central trough modelled in the mud plaster, and smaller gullies from around the base of the bench ran into the trough; the purpose or function of these features is unfathomable. There were the sockets for a number of substantial wooden posts around the perimeter wall of the building, but they could not have supported a roof across such a large space. Whatever the building's function, it was carefully maintained over a period of centuries (Figure 6.3).

Adjacent to the exterior wall there was a series of small circular or oval structures that the excavators have very carefully shown to have been for bulk storage of crops that needed to be kept dry. And the evidence of the many pestles and mortars, the use of chaff in the pisé walls of buildings, and the carbonised grains of wild barley suggest that the inhabitants were practising cultivation, that is the cultivation of a wild species. The age profiles of the wild goats that provided the main source of meat suggest that they were being hunted in a selective manner to manage the

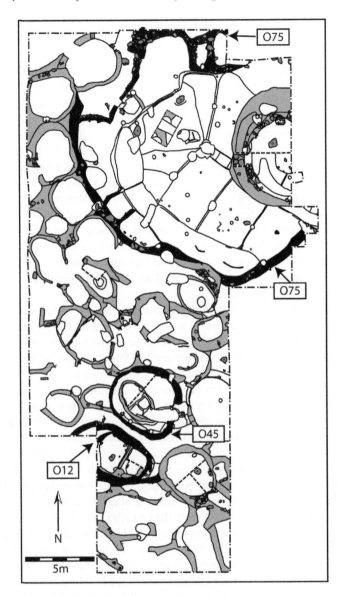

Figure 6.2 WF16 in southern Jordan, plan of the excavated area. The large communal building (labelled 075) and the two probable granary structures are shown in black.
(A later building overlays part of the large communal building.) (Finlayson et al.
2011, Fig. 2, with permission)

herds. As has been found at a number of other sites of the early Pre-Pottery Neolithic, foxes were hunted, presumably for their fur, and the bones of raptors suggest
that their feathers and talons were prized for their symbolic value.

Figure 6.3 WF16, the large communal building 075. The 'bench', some of the post sockets, and the radiating runnels shaped in mud plaster can be seen. (Finlayson et al. 2011, Fig. 3, with permission)

Over the decades since Kathleen Kenyon was surprised by the numerous human bodies buried within the Pre-Pottery Neolithic settlement of ancient Jericho, many more Pre-Pottery Neolithic settlements have produced numbers of intramural burials, some with missing skulls. The excavators of WF16 were able to learn from those decades of experience as they examined the many burials that had been inserted under the floors or cut into the walls of the buildings. They have found that in some cases bones had been removed from or added to burials. They also found collections of bones that had been painted and wrapped in plaster bundles. There was not a single common practice for the treatment of the dead: while some bodies were simply interred under a floor, the collections of bones indicate that other bodies had been somewhere else for some time before the bones were collected, and some burials had been 'revisited' for the removal or addition of bones.

When we shift our focus to the north Levant, the sites that have been investigated are all, with two important exceptions, salvage excavations ahead of flooding by major dams in the Euphrates and Tigris valleys both in north Syria and in southeast Turkey. Two in the Euphrates valley in north Syria and two or three more in the upper Tigris valley in southeast Turkey were founded either just before or during the Younger Dryas phase, suggesting that the environment was not as severely impacted across the northern arc of the hilly flanks zone as was argued some years ago (Hillman et al. 2001). Other sites date within the early Pre-Pottery Neolithic, sometimes from its very start around 9600 BCE. Wherever their Epipalaeolithic ancestors were living, it was not in the great river valleys, until almost the end of the Epipalaeolithic period. We can conclude that there was something in the new way of life represented by the earliest Pre-Pottery Neolithic communities that made the river valleys magnetic. One thing that we know was new about this period was the gradually increasing amount of pre-domestic cultivation of cereals.

The construction of a massive dam in the early 1970s on the bend of the Euphrates about 100 km east of Aleppo in north Syria brought an international army of archaeologists to undertake urgent salvage excavations on the many sites of all periods that were threatened with destruction by drowning. In many ways, the Euphrates valley in north Syria was *terra incognita* for archaeologists, and there were many discoveries that were completely unanticipated. Among them were two that we met in the previous chapter, because they both originated in the final centuries of the Epipalaeolithic period. Mureybet was a small tell-site that was occupied continuously for more than 2000 years, from the end of the Epipalaeolithic, through the early Pre-Pottery Neolithic and into the beginning of the later Pre-Pottery Neolithic. It was excavated between 1971 and 1974 by a French team led by Jacques Cauvin (Ibañez 2008a, 2008b, which is a convenient summary of the conclusions). The second site was on the opposite bank, only a few kilometres downstream, and it lay under the still occupied village of Abu Hureyra (Moore et al. 2000). When they found that the earliest occupation of the site was at the end of the Epipalaeolithic period, the trio who managed the excavation and research, Andrew Moore (archaeologist), Gordon Hillman (archaeo-botanist) and Tony Legge (archaeo-zoologist), hoped to have a continuous sequence of occupation that would document the process of cultivation and animal management leading to the domestication of plants and animals and the developing of a farming economy. But there was a substantial hiatus between the end of the Epipalaeolithic occupation and re-occupation of the same site in the later Pre-Pottery Neolithic, which will find its place in the next chapter.

At the base of the Tell Mureybet stratigraphy there was a thin stratum dating to the final part of the late Epipalaeolithic period, when there were circular semi-subterranean houses like those in the southern Levant, and a chipped stone industry that was practically indistinguishable from the classic Natufian. That was overlain by a succession of strata where the material culture transitioned rapidly out of the typical Epipalaeolithic microliths, replaced by the first types of small projectile points. The first projectile point to appear is called the Khiam point (named after the site in modern Israel where it was first identified in the middle of the last century). In the succeeding strata the settlement's architecture became more solid and more varied, and more styles of making projectile point appeared. Within the quite small area of the excavations there were more of the small, circular, semi-subterranean buildings, but there were others that were built on the ground surface, some with a foundation course of stone, others with walls made entirely of mud, and others again that were constructed of cigar-shaped lumps of soft limestone laid in mud mortar. There were also a few rectangular buildings internally sub-divided into two or four small rooms. One circular building was somewhat larger than the

Figure 6.4 Jerf el Ahmar. The second of the series of communal buildings, with its symmetrical pattern of doorless cells opposite mud-plaster platforms. In the centre
of the floor is the spreadeagled body of a young, headless female. (By courtesy
of Danielle Stordeur)

others, about 6 m in diameter. Only part of the building lay within the excavation
trenches, but that part of it was sub-divided into a series of cell-like rooms. It is the
first of such buildings to have been found, and when we move upstream to the site
of Jerf el Ahmar we shall see complete examples that have been found there and at
other settlements higher up the Euphrates valley (Figure 6.4).

More dam-building on the Euphrates followed both further upstream in north
Syria and in southeast Turkey. A cluster of three small settlements of early Pre-
Pottery Neolithic date were briefly investigated alongside the Euphrates in Syria,
close to the border with Turkey. The site of Jerf el Ahmar saw the most extensive
excavations, which were led by Danielle Stordeur, who had worked with Jacques
Cauvin at Mureybet, and had taken over the leadership of his research group when
he retired. Jerf el Ahmar was occupied only within the early Pre-Pottery Neolithic
period, which meant that it could be explored extensively (Stordeur 2000, 2015).
The settlement was situated on twin natural rises on the terrace edge overlooking
the Euphrates and its narrow floodplain. Over the centuries of its life, the settlement
had shifted about, first occupying one of the rises, then the other. The radiocarbon
dates show that the settlement was founded close to the beginning of the early Pre-
Pottery Neolithic and the occupation ended as the early Pre-Pottery Neolithic was
turning over into the late Pre-Pottery Neolithic, a settlement history of about 800
years (Figure 6.5).

The houses were simple, small, roughly circular buildings, as at all of the other
early Pre-Pottery Neolithic settlements. They were not fitted with hearths or ovens,
but there were substantial outdoor hearths each of which seems to have been shared

Figure 6.5 Jerf el Ahmar. The next communal building had a mud-built 'bench' with large
'kerb-stones' set between the main roof support posts. (By courtesy of Danielle
Stordeur)

by a cluster of houses. The houses were situated along terraces formed by stone-built walls. That surely implies that settlement was planned, and the terrace walls were laid out and constructed (by the community) before the houses were built. The most remarkable feature of Jerf el Ahmar is the series of at least four, large, subterranean communal buildings. The best preserved is the second in the series (its predecessor was very similar in plan, but was less well preserved and could not be completely excavated before the site was flooded by the rising waters behind the completed dam) (Figure 6.6).

The second communal building was a massive subterranean construction at the centre of a cluster of buildings that were different from the regular houses. It was 7 m in diameter and dug more than 2 m deep into the ground (requiring the removal of some 77 m² of material). It was built in its cylindrical cavity with a wall of stones set in mud mortar, and it had been completely rebuilt at least once (the first stone wall was completely replaced by a second). The roof had been supported on a series of wooden posts that were set into the stone wall, and the whole interior had been finished with mud plaster surfaces. The interior of the building consisted of a symmetrical set of five doorless cells around two thirds of the circle; the rest of the interior was open and consisted of three rammed earth platforms. In size and plan it was remarkably similar to a building that had been partially within the excavation area at Tell Mureybet. At the end of its life, the building was deliberately destroyed and the space that it had occupied was filled in (again, involving many cubic metres of soil). In the dramatic staging of its destruction, the first act was to throw through the trapdoor in the flat roof the body of a young woman, whose skeleton was found

Figure 6.6 Jerf el Ahmar. The kerb of the 'bench' in the communal building (Figure 6.5). (By courtesy of Danielle Stordeur)

spreadeagled in the centre of the floor. The wooden roof support posts were pulled out, and the collapsed roof timbers and other material was set on fire, which burned the body on the floor so that the bones were found reddened and blackened. The skeleton was headless, but its removal was so neat – no cut-marks on cervical vertebrae – that the body must have been completely decomposed or consumed in the fire; the implication is that the cranium and mandible had been carefully retrieved after the fire. Two or three crania were found in the bottom of post-sockets in other communal buildings (but none of them belonged to the female whose skeleton was found). The five doorless cells were huge storage bins, but the function of the carefully made floor levels in the open part of the building is unknowable. The scale of the construction, the careful rebuilding of the structure, and the very deliberate ceremonies with which its life was ended are all tokens of the importance of the building (and the sequence of other similar communal buildings) to the whole community of Jerf el Ahmar.

Fortunately for the excavators, the fire that consumed the roof had carbonised the residues of wheat, barley and lentils that had been stored in the cells, showing that the building had served in large part as a storage facility large enough to serve the whole community. Around the communal storage building were several communal kitchen buildings each equipped with multiple grinding stones. In several different ways Jerf el Ahmar shows us a small community (Danielle Stordeur has estimated its population around 150 persons) that invested greatly in communal effort. As we saw in Chapter 2, the team's archaeo-botanist George Willcox was able to plot the gradual changes in the barley that the community had harvested over many centuries, showing that it was slowly changing in size towards the domesticated form. Although the community were engaged in what he has labelled pre-domestic cultivation, they were not farmers. Nevertheless, the amounts of wild harvests they were gathering into their bins was impressive. And they were continuing to behave rather in the sharing ethos typical of hunter-gatherers.

For many years we had the idea that the first settlements of the early Pre-Pottery Neolithic period were simply small clusters of simple houses; the period had been labelled by Robert Braidwood as the emergence of 'village-farming'. That idea was false from the start, as Kenyon's Jericho excavations had shown; Pre-Pottery Neolithic Jericho, with its circuit wall, rock-cut ditch, great circular tower and intramural burials was not some simple village. Since the 1980s there has been a succession of unexpected and extraordinary discoveries of early Pre-Pottery Neolithic settlements, like Jerf el Ahmar, that were found within salvage archaeology projects (that is, not in consequence of deliberate archaeological research plans). An early example was the small site of Hallan Çemi, on a tributary of the Tigris, near the city of Batman (Rosenberg 2011). A series of accelerator mass spectrometer radiocarbon dates places its occupation over a few centuries at the very beginning of the early Pre-Pottery Neolithic. There were some small, circular houses, but, in a later phase, there were also two larger, more substantially built, semi-subterranean buildings beside an extensive open area at the centre of what was clearly a deliberately laid-out settlement (Figure 6.7). On the basis of the dense concentration of fire-cracked stones and articulated limbs of large game animals and whole skulls

Figure 6.7 Jerf el Ahmar. One of the multi-roomed rectangular buildings beside a communal building. One of the rooms was equipped with multiple quern-stones for grinding grains and seeds. (By courtesy of Danielle Stordeur)

of sheep and goats, Rosenberg identified this 15-m diameter area as the 'central activity area', the focus of community feasts (Rosenberg et al. 1998; Rosenberg & Redding 2000).

The massive skull of a wild bull with broad horns was found at the foot of the wall of one of the two large buildings (Figure 6.8); it lay where it had presumably fallen from its attachment to the wall. Rosenberg suggested that these were 'public buildings' (similar to Danielle Stordeur's labelling of 'communal buildings' at Jerf el Ahmar). Both the plant remains that were recovered and the faunal assemblage

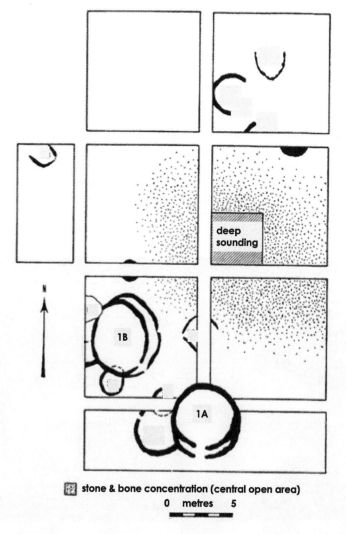

stone & bone concentration (central open area)

0   metres   5

Figure 6.8 Hallan Çemi plan. The two large, circular, semi-subterranean buildings had each been rebuilt at least once. They stood beside an open area whose surface was densely scattered with fire-cracked stones and animal bone. (By courtesy of Michael Rosenberg)

are significant. The plant remains were recovered by flotation and first studied by Mark Nesbitt (who had worked with me in the late 1980s at the site of Qermez Dere in north Iraq, of which more below). There was a conspicuous absence of cereals; instead there were various grasses, sea-club rushes, almonds, pistachio, lentils and vetches (Savard et al. 2006). The recovered faunal material amounted to more than 2 metric tons, which has supplied more than enough for a series of detailed studies that are summarised by Zeder and Spitzer (2016), who were particularly concerned with the quantity and diversity of bird bones. It might be thought that the choice of Hallan Çemi in the piedmont of eastern Turkey for a new settlement at the end of the Younger Dryas phase and the first centuries of the Holocene was odd, but the animal bones include plenty from medium-sized species as well as representing a 'broad spectrum' of species. The bones demonstrate that the site sat at a strategic location at the junction of different environmental zones: the people of Hallan Çemi knew very well what they were doing and how to exploit the seasonality of the varied local habitats. Finally, Zeder and Spitzer (2016) show that there were differences between everyday life among the small houses and the communal events in the 'central activity area', where there were the remains of large joints of the larger hunted species, and significant differences in the representation of species of birds and their body parts. In the 'central activity area' the larger bird species were over-represented, and particularly the large raptors such as eagles and vultures, which were most likely prized for their symbolic significance. Similar observations have been made on the bird bones at a number of late Epipalaeolithic and early Pre-Pottery Neolithic sites (Dobney 2002), notably at Jerf el Ahmar and Tell Mureybet (Gourichon 2002; Gourichon & Helmer 2008) (Figure 6.9).

Figure 6.9 Hallan Çemi. The skull of a wild bull had fallen from the wall of one of the large circular buildings. (By courtesy of Michael Rosenberg)

For a while, Hallan Çemi was an isolated discovery, but in recent years there has been a string of four early Pre-Pottery Neolithic settlements found along the Tigris valley. Their excavation has been part of the salvage operations ahead of the completion of a major dam that is flooding more than 100 km of the narrow valley. Excavations are still in progress at two of them, and not much information available at present. On the one hand these four early Pre-Pottery Neolithic settlements share basic cultural characteristics; but on the other hand each is different from the others in certain particulars. They are small, and they have simple, small, circular houses; they share a tradition of chipped stone tool-making that is an extension into the early Holocene of the typical Epipalaeolithic bladelet industries (Maeda 2018); they each have semi-subterranean buildings, and three of them share the characteristic feature of having special buildings with pairs of upright stone pillars. Two of the settlements were founded in the middle of the Younger Dryas phase, and the occupations at the other two seem to begin early in the early Pre-Pottery Neolithic.

Each community had a distinctive tradition for the burial of the dead within the settlement. Körtiktepe (of which more below) is made up of densely packed semi-subterranean circular houses, and more than 890 burials have been found, mostly below the floors of houses. At the other settlements in this little group there are some, or few, or almost no burials. The group of bio-archaeologists who have examined the plant and animal remains excavated at Gusir Höyük have remarked that these early Pre-Pottery Neolithic communities 'pursued highly idiosyncratic plant and animal exploitation practices targeting plant "staples" that diverged markedly even between sites located in close proximity to each other' (Kabukcu et al. 2021). Another group who sourced the obsidian found at these sites noted that amounts varied considerably from site to site (Carter et al. 2021). At Hallan Çemi, which is located on the obvious route through the mountains to the east Anatolian obsidian sources, 58% of the chipped stone is obsidian. At Körtiktepe, near the confluence of that tributary with the Tigris, there is 45% obsidian. The other sites, downstream from Körtiktepe, have a mere 5%–9% obsidian, suggesting that access to the raw material played a part as well as community-specific practices concerning the use of preferred raw materials. In sum, while these settlements shared much of their cultural traditions and were clearly in touch with another, as evidenced by the materials and things that they exchanged, in certain elements of community life, particularly regarding community buildings and traditions of intramural burial, each settlement followed its own distinct version of the common cultural practices.

In addition to its numerous burials there is another good reason for mentioning the site of Körtiktepe (Özkaya & Coşkun 2011), which is one of those whose initial occupation dates from the Epipalaeolithic-Younger Dryas period (Benz et al. 2015). Many of the numerous burials at Körtiktepe were found to include large quantities of artefacts, especially beautifully decorated carved stone bowls. Similar bowls, or fragments of bowls, have been found at other sites in southeast Anatolia, and in the Euphrates valley in north Syria (Figure 6.10); but at Körtiktepe they are present in profusion, perhaps because they were being made there from a local stone. The stylised devices that decorate the surface of the bowls have been found repeated on other small stone artefacts, and they also occur on a much larger

Figure 6.10 Hallan Çemi. One of the large, semi-subterranean buildings. The recesses in the stone retaining wall accommodated the wooden posts that supported the roof. (By courtesy of Michael Rosenberg)

Figure 6.11 Decorated chlorite bowl from Jerf el Ahmar. (By courtesy of Danielle Stordeur)

scale on some of the massive T-shaped monoliths of Göbekli Tepe (to which we will shortly turn) (Figure 6.11).

Several of the settlements along the Tigris valley have produced buildings that are semi-subterranean, and that possess pairs of tall, stone pillars, sometimes one

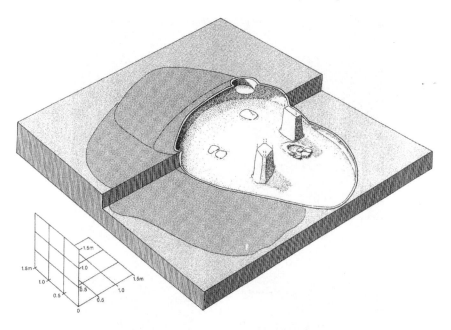

Figure 6.12 Qermez Dere: the third rebuild of a semi-subterranean house with a pair of clay-covered stone pillars.

pair, sometimes two pairs, set upright in their floors (Figure 6.12). In some cases, the semi-subterranean building is distinctly larger and more solidly constructed than the houses, making it a candidate for being some kind of communal building. But in other cases, buildings that otherwise appear the same as the houses are furnished with a pair of pillars. I am immediately reminded of our experience with the salvage excavations at the little early Pre-Pottery Neolithic settlement of Qermez Dere on the outskirts of the town of Telafar in northern Iraq (Watkins 1990). At that time, having to decide how to describe a novel kind of building that was furnished with pairs of pillars that were modelled in clay around stone core, and lacking any other kind of building in the settlement, I decided to call them houses. They were, I suggested, houses that accommodated objects with symbolic meaning, in rather the same way that my house today is made into a 'home' with the display of wedding and family photos, or a piece of vintage furniture that carries special memories. In spite of roughly 30 years of experience of these various kinds of early Neolithic buildings, we are not yet clear about how we should interpret them. The massive and very large examples are obviously special-purpose buildings, but we now have a growing number of ambiguous buildings (Figure 6.13).

The archetype of monumental circular buildings are those that have been excavated at Göbekli Tepe, near Urfa in southeast Anatolia (Schmidt 2011). Göbekli Tepe is the location for the most massive, dramatic and expressive of circular structures, with the most elaborately decorated T-shaped monoliths, and other carved imagery. But they are unique, and there is no clear and accepted interpretation of

Figure 6.13 Göbekli Tepe. One of the central pair of T-shaped monoliths in special building D. It is 5.5 m tall, standing in a plinth carved from the bedrock. The figure
wears a belt with an elaborate buckle, from which hangs a fox-skin. At its neck
the figure has a collar on which there is a pendant carved with symbols. The
other monolith has a similar belt, buckle, fox-skin, collar and pendant, but the
pendant has different symbols. (By kind permission of the DAI)Göbekli Tepe.
One of the central pair of T-shaped monoliths in special building D. It is 5.5 m
tall, standing in a plinth carved from the bedrock.

the symbolism of the architecture or the sculptured stones. (Other sites in the region
around Urfa and Göbekli Tepe are now under excavation and are also beginning to
produce large, circular special buildings and more T-monoliths.) We do not know
enough, even after 30 years of investigation, to give an archaeological interpretation; and that has left a vacuum that has been filled with excited speculation about
the buildings as 'the first temples' and the anthropomorphic T-shaped monoliths as

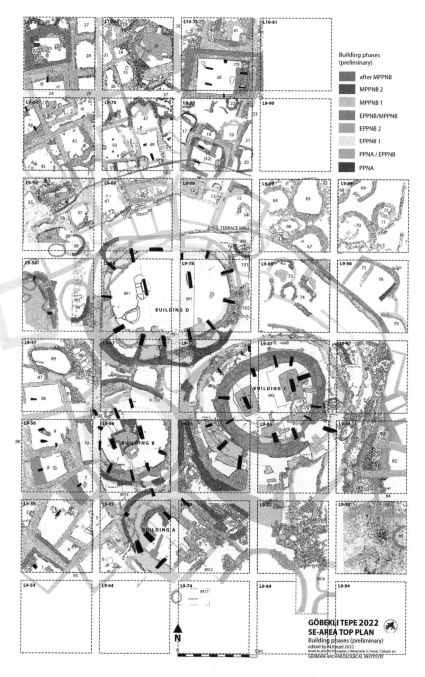

Building phases
(preliminary)

after MPPNB
MPPNB 2
MPPNB 1
EPPNB/MPPNB
EPPNB 2
EPPNB 1
PPNA / EPPNB
PPNA

GÖBEKLI TEPE 2022
SE-AREA TOP PLAN
Building phases (preliminary)
edited by M.Kinzel 2022
based on plans by D.Johnphat, C.Winterstein, K.Fischer, T.Götzelt, etc.
GERMAN ARCHAEOLOGICAL INSTITUTE

Figure 6.14 Göbekli Tepe. Plan of the cluster of four special buildings at the southeast edge
of the mound. Each has a pair of central T-monoliths, and more monoliths are
set around the perimeter. There are clear signs that each special building was
rebuilt and repaired, and the radiocarbon dates show that they existed at the
same time, and over a long period. (By kind permission of the DAI)

'the first gods'. The site of Göbekli Tepe is much more than the cluster of four large circular buildings whose careful excavation has occupied most of the excavation work; it is a very large site, about 300 m in diameter and as much as 15 m thick with accumulated archaeological deposits. We still know very little about the nature of the mounded settlement beyond the monumental circular buildings; and the location of Göbekli Tepe, on a bare, rocky mountain ridge, where there is almost no soil, and no water, contributes further to its mystery. Klaus Schmidt did publish a general outline book, but I fear it was an error to title it: 'Sie bauten die ersten Tempel' – they built the first temples (Klaus Schmidt 2006). In recent years, much attention has been given to creating protective roof structures over the excavated areas and providing visitor facilities; an update on the German Archaeological Institute's recent research shows that our understanding of the site is moving forward in ways that differ from earlier assumptions (Clare 2020) (Figure 6.14).

The four large, circular buildings (that are still not completely investigated) form a tight cluster near the southeast edge of the mound. Their massive walls were built of stones set in mortar made of mud tempered with chaff (from which samples were obtained for radiocarbon dating). At least two of the buildings have two or even three concentric walls; it appears that buildings that were as much as 30 m in diameter were reshaped once or twice on a smaller scale. Nevertheless, they are 20 m in diameter. One of the buildings has an entrance passage leading to a narrow doorway, but the others have no doorways at floor level; it has to be assumed that they were accessed by ladders leading down from trapdoors in their roofs. (which certainly seems to be the case with the communal buildings in the middle of the settlement of Jerf el Ahmar, in north Syria). Around the base of the wall of each building there is a stone-built bench-like structure; and around the perimeter there is a series of about ten or twelve tall stone monoliths that are set into the bench and into the wall. Like the pair of central monoliths that each building has, the monoliths are shaped into a T-form, and they have quite sharply defined edges and smoothed surfaces. The pair of central monoliths are taller than those around the walls. The tallest pair of central monoliths are fully 5.5 m tall. Like several other monoliths, the tallest pair have sculpted arms and hands; they also wear a belt, from which hangs what appears to be a fox-skin loincloth. And each wears a collar around his neck from which hangs a pendant. Each of the many T-monoliths that have been recorded has motifs on the two sides; some of the motifs are wild, usually male, animals, but there are also birds, reptiles, outsize spiders, scorpions and many snakes (Figure 6.15).

The excavators have found evidence that the four buildings were repaired and restored from time to time. The radiocarbon dates from the chaff that was incorporated into the mortar of the walls show that the buildings were in use for several hundred years, and were still in use into the later Pre-Pottery Neolithic period. They were not built in sequence, with one replacing another; but rather they coexisted. And there are a number more such massive circular structures in other parts of the site, most of them known from geophysical survey. We know that the cluster of four buildings were, in the end, filled with hundreds of tons of soil and stones, leaving only the top edges of the tallest monoliths visible; and we also know that, at the top of the mound, overlooking the area where the four big

Figure 6.15 Small stone plaquette with incised signs. Such small stone plaquettes, carved
with symbols, such as snakes, that are also found on the T-monoliths of Göbekli
Tepe, have also been found at other settlements in the region, such as Jerf el
Ahmar. (By kind permission of the DAI).

buildings had stood, there were densely packed rectangular, house-like buildings
that date to the early part of the later Pre-Pottery Neolithic period. A number of
these smaller rectangular buildings also contain pairs of smaller T-monoliths. One
final observation: it is clear that the task of constructing, or reconstructing, one of
the massive buildings, quarrying the monoliths from the bare rock nearby, carving
and erecting the monoliths would have required a very large labour force (which
itself would have needed substantial logistical support with food and water); on the
other hand, each massive building would not have accommodated a large number
of people (and the T-monoliths suggest that around ten or a dozen people met in a
group around a central pair of taller – more important? – persons). If we could only
decode the significance of the sculpted monoliths, perhaps we could say what kind
of special sub-group they were who met together out of sight of the community.

   Before moving on from Göbekli Tepe it is important to note that there are
now excavations in progress at one of several more sites of the early Pre-
Pottery Neolithic period within the region around the city of Urfa. Karahantepe
has already produced a massive circular semi-subterranean building similar to
those of Göbekli Tepe (Karul 2021). And we also know that there are many
T-monoliths. At present part of one large (23 m diameter) sub-circular, sub-
rectangular structure has been excavated. Unlike those at Göbekli Tepe, this
structure was cut into the bedrock of the hill on which the site is located. There
have been differences of opinion concerning the fill of the buildings at Göbekli
Tepe; Schmidt and his team were sure that the buildings had been deliberately
closed, their roofs removed, and the whole space deliberately filled, but others
have proposed that the fill was the product of erosion from the crown of the
settlement mound once the site was abandoned. The excavator of Karahantepe
is quite clear that the communal building there was deliberately backfilled
(Karul 2021). We were sure that the subterranean buildings at Qermez Dere

were deliberately buried, and immediately replaced; and Stordeur was also sure that the series of communal buildings at Jerf el Ahmar were burnt and buried at the end of their lives.

To emphasise the point that large, circular, communal buildings are not exclusive to one part of the hilly flanks zone, we should note that the recent to the site of Tell Asiab, in a Zagros valley in western Iran, has produced another example (Richter et al. 2021). Robert Braidwood's team excavated briefly at Tell Asiab in 1960: the new excavations set out to expand the area of the earlier trench, in which a small part of the curvilinear outline of a building was found. We now have much more of the building, amounting to about half of its perimeter wall, and we can see that it was around 10 m in diameter, and was cut approximately 1.2 m into the natural subsoil. A pit in the floor was found to contain the crania and mandibles of 19 wild boar, the crania stacked on one side and the mandibles arranged neatly on the opposite side of the pit. Centrally and beneath the wild boar remains was the skull of a brown bear. This is one of several similar deposits elsewhere that seem to be the ceremonial marking of a major feast (Banksgaard et al. 2019). A series of new radiocarbon dates place the occupation of the site to the middle of the tenth millennium BCE, that is early in the early Pre-Pottery Neolithic period.

At the end of the previous chapter on the late Epipalaeolithic period we saw that there were significant sites beyond the arc of the hilly flanks zone in central southern Anatolia (Pınarbaşı) and Cyprus (Aetokremnos). Now we can add in this chapter that there were important settlements of the early Pre-Pottery Neolithic, in both those regions. There is a chronological gap (equivalent to more than the millennium of the Younger Dryas) between the burials in the rock-shelter at Pınarbaşı and the establishment of a small cluster of semi-subterranean houses a couple of hundred metres away on the edge of a pool and the great expanse of seasonal marshland (Baird 2012; Baird et al. 2011). The settlement is made up of a cluster of circular semi-subterranean houses, with plastered surfaces on their floors and lower walls, and probably wattle and daub superstructures. There were burials of both complete bodies and the partial remains of others within the settlement. Today it is an isolated place that is difficult to access, but the early Neolithic community belonged within a social network, obtaining marine shells from the Mediterranean and using obsidian from the sources more than 100 km distant to the east. The group who had chosen to settle at Pınarbaşı exploited a unique environment. The faunal remains show that they hunted wild aurochs, equids, and some sheep, goat and deer out on the plains and the hills, and they were expert wild-fowlers and fishers, exploiting the pool and the wetlands that began only metres from their homes. They collected wild almonds and terebinth, and probably some plants from the wetlands, but they were not well placed for any exploitation of wild cereals. The Pınarbaşı settlement began at the very beginning of the early and continued into the beginning of the later Pre-Pottery Neolithic periods.

In the previous chapter we saw that the first people to colonise the island of Cyprus were hunting wild boar which they had brought from the mainland to populate the woodlands. That first chapter in the island's human history is dated to the boundary between the Younger Dryas and the Holocene. In the early Pre-Pottery Neolithic period there is a gap, but whether that is a gap in our knowledge or an actual break in the human occupation of the island we do not know. Until recently, the next evidence of human occupation was a series of further colonisations in the later Pre-Pottery Neolithic, but two sites have been discovered that take those colonisations back into the end of the early Pre-Pottery Neolithic period, in the early centuries of the ninth millennium BCE. One of them, Klimonas, was a well-established community of more than 50 circular houses that is mostly known from geophysical survey (Vigne 2017; Vigne et al. 2012). The one house that has been partly excavated was about 6 m in diameter, and was cut into the slope of the hill-side; it had a pisé wall, a neatly plastered floor, internal posts supporting the roof, and pits that concealed what were presumably significant artefacts. By contrast, Structure 10 was a semi-subterranean circular building that was fully 10 m in diameter, and is another example of a communal building. The roof was supported by a large central post and a series of pairs of lighter posts around its perimeter. In the interior of the building there were several hearths, pits and low peripheral benches with rectilinear designs on their edges. This communal building had been restored and reconstructed on at least four occasions, each time being given a new floor. The settlement and its communal building had lasted for several centuries. Careful analysis of the mud that was used for the walls and floors of the buildings indicates that the people of Klimonas villagers were cultivating cereals, especially emmer wheat, which must have been introduced from the mainland. The animal bones show that they were intensively hunting the island's wild boar that had been introduced in the late Epipalaeolithic. They possessed domestic dogs that were probably used in hunting, and they also had domestic cats that no doubt helped to keep down the accidentally introduced house-mouse population.

The information that we have of these early Cypriot colonists may be rather thin, but it is important for what it tells us about population and population density on the nearest part of the southwest Asian mainland, the north of the Levant. The community that established themselves at Klimonas were clearly not seasonal visitors to Cyprus, keen to enjoy some trophy-hunting of wild boar: they constituted a sizeable and permanent community that had brought a cultural package from the north Levant. It seems to me that they are evidence that levels of population density on the mainland in the middle of the early Pre-Pottery Neolithic were such that some groups found it necessary to plan and establish a new home in Cyprus. In short, from the early Pre-Pottery Neolithic period, if not from the end of the Epipalaeolithic (recall Aetokremnos in the previous chapter), there was sufficient population pressure within at least some parts of the hilly flanks of the Fertile Crescent to cause some groups to look for new lands into which they could expand.

References

Baird, D. (2012). Pınarbaşı: From Epipalaeolithic Camp-Site to Sedentarising Village in Central Anatolia. In M. Özdoğan, N. Başgelen, & P. Kuniholm (Eds.), *The Neolithic in Turkey. New Excavations and New Research: Central Turkey* (pp. 181–218). Istanbul: Arkeoloji ve Sanat Yayinlari.

Baird, D., Carruthers, D., Fairbairn, A., & Pearson, J. (2011). Ritual in the Landscape: Evidence from Pınarbası in the Seventh-Millennium cal BC Konya Plain. *Antiquity*, 85(328), 380–94. doi:10.1017/S0003598X0006782X

Banksgaard, P., Yeomans, L., Darabi, H., Gregersen, K. M., Olsaen, J., Richter, T., & Mortensen, P. (2019). Feasting on Wild Boar in the Early Neolithic. Evidence from an 11,400-Year-Old Placed Deposit at Tappeh Asiab, Central Zagros. *Cambridge Archaeological Journal*, 29(3), 443–63.

Bar-Yosef, O., Gopher, A., Tchernov, E., & Kislev, M. E. (1991). Netiv Hagdud: An Early Neolithic Village Site in the Jordan Valley. *Journal of Field Archaeology*, 18(4), 405–24. doi:10.1179/009346991791549077

Benz, M., Deckers, K., Rössner, C., Alexandrovskiy, A., Pustovoytov, K., Scheeres, M., . . . Özkaya, V. (2015). Prelude to Village Life: Environmental Data and Building Traditions of the Epipalaeolithic Settlement at Körtik Tepe, Southeastern Turkey. *Paléorient*, 41(2), 9–30.

Carter, T., Moir, R., Wong, T., Campeau, K., Miyake, Y., & Maeda, O. (2021). Hunter-Fisher-Gatherer River Transportation: Insights from Sourcing the Obsidian of Hasankeyf Höyük, a Pre-Pottery Neolithic A Village on the Upper Tigris (SE Turkey). *Quaternary International*, 574, 27–42. doi:10.1016/j.quaint.2020.09.045

Clare, L. (2020). Gobekli Tepe, Turkey. A Brief Summary of Research at a New World Heritage Site (2015–2019). *E-Forschungsberichte Des DAI*, 2, 1–13. doi:10.34780/efb.v0i2.1012 Retrieved (15 December 2022) from: https://publications.dainst.org/journals/efb/article/view/2596

Dobney, K. (2002). Flying a Kite at the End of the Ice Age: The Possible Significance of Raptor Remains from Proto- and Early Neolithic Sites of the Middle East. In H. Buitenhuis, A. M. Choyke, M. Mashkour, & F. Poplin (Eds.), *Archaeozoology of the Near East IV. Proceedings of 4th International Symposium on the Archaeozoology of Southwestern Asia and Adjacent Areas* (Vol. 32, ARC Publications, pp. 74–84). Groningen: Centre for Archaeological Research and Consultancy.

Finlayson, B., Mithen, S. J., Najjar, M., Smith, S., Maričević, D., Pankhurst, N., & Yeomans, L. (2011). Architecture, Sedentism, and Social Complexity at Pre-Pottery Neolithic A WF16, Southern Jordan. *Proceedings of the National Academy of Sciences*, 108(20), 8183–8. doi:10.1073/pnas.1017642108

Gourichon, L. (2002). Bird Remains from Jerf el Ahmar, a PPNA Site in Northern Syria, with Special Reference to the Griffon Vulture. In H. Buitenhuis, A. Choyke, M. Mashkour, & A. H. Al-Shiyab (Eds.), *Archaeozoology of the Near East V. Proceedings of the 5th International Symposium on the Archaeology of Southwestern Asia and adjacent areas* (pp. 138–52). Groningen: ARC.

Gourichon, L., & Helmer, D. (2008). Étude archéozoologique de Mureybet. In J. J. Ibáñez (Ed.), *Le site Néolithique de Tell Mureybet (Syrie du Nord). En hommage à Jacques Cauvin* (pp. 115–228). Oxford: British Archaeological Reports, International Series 1843, Archaeopress.

Hillman, G. C., Hedges, R., Moore, A. M. T., Colledge, S., & Pettitt, P. (2001). New Evidence of Lateglacial Cereal Cultivation at Abu Hureyra on the Euphrates. *The Holocene*, 11(4), 383–93.

Ibañez, J. J. (2008a). Conclusion. In J. J. Ibañez (Ed.), *Le site néolithique de Tell Mureybet: en hommage à Jacques Cauvin* (Vol. 2, pp. 661–75). Oxford: Archaeopress.

Ibañez, J. J. (2008b). *Le site néolithique de Tell Mureybet: en hommage à Jacques Cauvin (2 vols)* (Vol. British Archaeological Reports, International Series 1843). Oxford: Archaeopress.

Kabukcu, C., Asouti, E., Pöllath, N., Peters, J., & Karul, N. (2021). Pathways to Plant Domestication in Southeast Anatolia Based on New Data from Aceramic Neolithic Gusir Höyük. *Scientific Reports*, 11(1), 2112. doi:10.1038/s41598-021-81757-9

Karul, N. (2021). Buried Buildings at Pre Pottery Neolithic Karahantepe/Karahantepe Çanak-Çömleksiz Neolitik Dönem Gömü Yapıları 2021. *Türk Arkeoloji ve Etnografya Dergisi*, 86, 19–31.

Kenyon, K. (1957). *Digging up Jericho*. London: Ernest Benn.

Kenyon, K. M. (1960). *Archaeology in the Holy Land* (4th ed.). London: Benn.

Kenyon, K. M., & Holland, T. A. (1981). *Excavations at Jericho, Vol. 3, The Architecture and Stratigraphy of the Tell*. London: British School of Archaeology in Jerusalem.

Maeda, O. (2018). Lithic Analysis and the Transition to the Neolithic in the Upper Tigris Valley: Recent Excavations at Hasankeyf Hoyuk. *Antiquity*, 92(361), 56–73. doi:10.15184/aqy.2017.219

Mithen, S., Finlayson, B., Maričević, D., Smith, S., Jenkins, E., & Najjar, M. (2018). *WF16: Excavations at an Early Neolithic Settlement in Wadi Faynan, Southern Jordan: Stratigraphy, Chronology, Architecture and Burials*. London: Council for British Research in the Levant.

Moore, A. M. T., Hillman, G. C., & Legge, A. J. (2000). *Village on the Euphrates: From Foraging to Farming at Abu Hureyra*. Oxford: Oxford University Press.

Özkaya, V., & Coşkun, A. (2011). Körtik Tepe. In M. Özdoğan, N. Başgelen, & P. Kuniholm (Eds.), *The Neolithic in Turkey. New Excavations and New Research: The Tigris Basin* (pp. 89–127). Istanbul: Arkeoloji ve Sanat Yayinlari.

Richter, T., Darabi, H., Alibaigi, S., Arranz-Otaegui, A., Bangsgaard, P., Khosravi, S., . . . Yeomans, L. (2021). The Formation of Early Neolithic Communities in the Central Zagros: An 11,500 Year-Old Communal Structure at Asiab. *Oxford Journal of Archaeology*, 40(1), 2–22. doi.org/10.1111/ojoa.12213

Rosenberg, M. N. (2011). Hallan Çemi. In M. Özdoğan, N. Başgelen, & P. Kuniholm (Eds.), *The Neolithic in Turkey. New Excavations and New Research: The Tigris Basin* (pp. 61–78). İstanbul: Arkeoloji ve Sanat Yayinlari.

Rosenberg, M., Nesbitt, R. M., Redding, R., & Peasnall, B. L. (1998). Hallan Çemi, Pig Husbandry, and Post-Pleistocene Adaptations along the Taurus-Zagros Arc (Turkey). *Paléorient*, 24(1), 25–42.

Rosenberg, M., & Redding, R. (2000). Hallan Çemi and Early Village Organization in Eastern Anatolia. In I. Kuijt (Ed.), *Life in Neolithic Farming Communities: Social Organization, Identity, and Differentiation* (pp. 39–62). New York, NY: Kluwer Academic.

Savard, M., Nesbitt, M., & Jones, M. (2006). The role of wild grasses in subsistence and sedentism: new evidence from the northern Fertile Crescent. *World Archaeology*, 38(2), 179–196.

Schmidt, K. (2006). *Sie bauten die ersten Tempel. Das rätselhafte Heiligtum der Steinzeitjäger*. Munich: Beck.

Schmidt, K. (2011). Göbekli Tepe. In M. Özdoğan, N. Başgelen, & P. Kuniholm (Eds.), *The Neolithic in Turkey. New Excavations and New Research - The Euphrates Basin* (pp. 41–83). Istanbul: Arkeoloji ve Sanat Yayinlari.

Stordeur, D. (2000). New Discoveries in Architecture and Symbolism at Jerf el Ahmar (Syria), 1997–1999. *Neo-Lithics: A Newsletter of South-west Asian Lithics Research*, 1(00), 1–4.

Stordeur, D. (2015). *Le village de Jerf el Ahmar (Syrie, 9500–8700 av. J.-C.): L'architecture, miroir d'une société néolithique complexe*. Paris: CNRS Éditions.

Vigne, J.-D. (2017). Archaeozoological Techniques and Protocols for Elaborating Scenarios of Early Colonisation and Neolithisation of Cyprus. In A. Albarella, M. Rizzetto, H. Russ, K. Vickers, & S. Viner-Daniels (Eds.), *The Oxford Handbook of Zooarchaeology* (pp. 70–89). Oxford: Oxford University Press.

Vigne, J.-D., Briois, F., Zazzo, A., Willcox, G., Cucchi, T., Thiébault, S., . . . Guilaine, J. (2012). First Wave of Cultivators Spread to Cyprus at Least 10,600 y Ago. *Proceedings of the National Academy of Sciences*, 109(22), 8445–9. doi:10.1073/pnas.1201693109

Watkins, T. (1990). The Origins of House and Home? *World Archaeology*, 21(3), 336–47.

Zeder, M. A., & Spitzer, M. D. (2016). New Insights into Broad Spectrum Communities of the Early Holocene Near East: The Birds of Hallan Çemi. *Quaternary Science Reviews*, 151, 140–59. doi:10.1016/j.quascirev.2016.08.024

7   Late Pre-Pottery
Neolithic – Climax

We have reached the period for which we have most evidence, the greatest number of sites explored, some of the most dramatic discoveries, and the most data relating to the domestication of plants and animals and the early development of farming. The later Pre-Pottery Neolithic is the time when communities took the experiment of living together in large numbers as far as it was possible to go. Around the hilly flanks and in central Anatolia settlements exhibit a great degree of cultural diversity. It was a time of dynamic cultural change and development; and at its latter end it leads on to the beginning of another transformation, which will occupy us in the next chapter.

The later Pre-Pottery Neolithic period was first defined by Kathleen Kenyon at Jericho as the upper block of the Pre-Pottery Neolithic strata, which she labelled the PPNB. It was clearly distinguished from the PPNA strata below by the rectangular shape of the buildings, the shape of the mud bricks, and the characteristics of the chipped stone industry. Between the two blocks of strata there was a chronological hiatus, when the mound was unoccupied. Her excavation results are massively reported (Kenyon & Holland 1981, 1982, 1983), but a simple synthesis is found in her popular account of the excavations (Kenyon 1957). Kenyon went on to write a concise account of the prehistory of the southern Levant, using the tags PPNA and PPNB as labels for successive stages of the Pre-Pottery Neolithic (Kenyon 1960). The term PPNB thereby became the name of an archaeological culture-period, that is, both a cultural package that could be recognised at a number of sites across a region, and a block of prehistoric time characterised by those cultural practices.

Over the decades, as more and more sites came to be excavated, archaeologists noted subtle changes in tool-making techniques and fashions for particular tool-forms within the period. The later Pre-Pottery Neolithic (or the PPNB) is usually divided into three, or sometimes four, sub-periods. We saw the phenomenon of accelerating cultural change across the Epipalaeolithic by contrast with the earlier Palaeolithic, and the pace of cultural change across the end of the later Epipalaeolithic and through the early Pre-Pottery Neolithic was notably greater again. But the fact that archaeologists recognise three or four sub-periods within the later Pre-Pottery Neolithic on the basis of material culture change tells us that cultural change and cultural accumulation were accelerating at an unprecedented rate. Later it became clear that the cultural traits that characterise the later Pre-Pottery

DOI: 10.4324/9781351069281-8

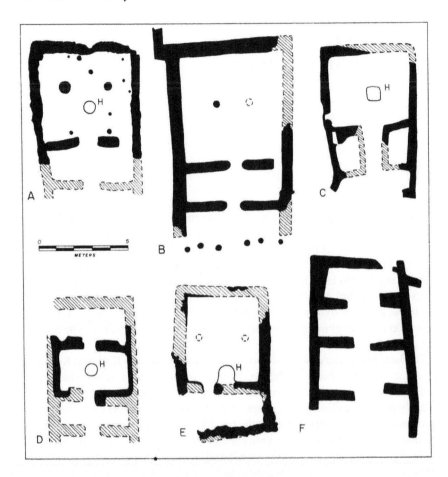

Figure 7.1 'Pier-houses' of the later Pre-Pottery Neolithic southern Levant. (Byrd &
Banning 1988, Fig. 2, with permission)

Neolithic appear in a staged transition from the earlier cultural traditions, at sites in
north Syria, particularly at Tell Mureybet. With the full range of the period charted,
archaeologists could refer to early, middle and late sub-periods (EPPNB, MPPNB
and LPPNB), and some have added a final sub-period, PPNC, that bridges the
transition out of the Pre-Pottery Neolithic and into the Late (Pottery) Neolithic.
For our purposes here, these refinements of terminology and internal chronology
are unnecessary, but we may note that the identification of these sub-periods by archaeologists tells us that cultural change within the late Pre-Pottery Neolithic was
happening at a faster rate than ever before (Figure 7.1).

Since Kenyon's discoveries at Jericho have become the standard for this period,
we should start there. The PPNB occupation at Jericho represents the re-use of
a site that had been abandoned for some centuries. The great wall that was first
erected in the early Pre-Pottery Neolithic was renewed. The new settlement was

Figure 7.2 This sub-floor burial at 'Ain Ghazal is typical of many at settlements of the later Pre-Pottery Neolithic. The burial has been re-opened and the skull has been retrieved, another common feature of these intramural burials. (By courtesy of Gary Rollefson)

established on a mound of earlier deposits that was already 4 m thick. The houses were built of cigar-shaped mud bricks laid in mud mortar on stone foundations. They were rectangular in form, though the corners were generally rounded rather than bonded and squared at right angles. In between the closely packed houses there seem to be irregular small open spaces or yards. In general the Jericho PPNB houses belong to what has been called the 'pier-house' tradition (Byrd & Banning 1988). Similar houses have been found at a number of sites in Israel, the Jordan valley and Jordan. They consist of two or three rooms along the long axis of the building, with doorways from one to the next room on the building's axis. There is often a hearth in the centre of the main room. Kenyon was very struck by the excellent lime plaster with which the interior surfaces of the walls and the floors were lined. Often the lime plaster was coloured red, and it was usually burnished. The use of lime plaster and red colouring is common at other sites of the period. At a number of settlements in the Levant of the later Pre-Pottery Neolithic the rectangular houses were of two storeys (Figure 7.2).

In amongst the houses of the settlement there were numerous burials. They were much more frequent than in the early Pre-Pottery Neolithic strata. Many of the graves contained multiple bodies, and there was sometimes evidence of disturbance that would suggest that a grave was used for successive burials. A significant number of the skeletons lacked skulls, and, conversely, Kenyon found clusters of skulls deposited in several houses. Some of the crania had their mandibles held in place with plastered facial features. The plaster might be painted, and in some examples the eyes were formed with inlaid cowrie shells. The plastered skulls of Jericho have become iconic, and illustrations are found everywhere.

Once we leave Jericho, the later Pre-Pottery Neolithic period becomes complicated: at one level there are cultural commonalities, such as the traditions of working chipped stone and the tools and projectile points that were produced, but at other levels each settlement, each community, seems to have its own distinctive ways, variants on a common theme. Before we set out to look at the diversity of the settlements, we should spend a minute with the working of chipped stone because it raises some points of considerable interest for us.

From the start of the later Pre-Pottery Neolithic period a new mode of chipped stone technology emerged around the Levant and the northern arc of the hilly flanks zone. As in the early Pre-Pottery Neolithic people continued to make and use large numbers of projectile points, and it is the changing fashions for the design of those projectile points that has allowed archaeologists to define the sub-periods within the later Pre-Pottery Neolithic. Stefan Kozlowski has suggested that we throw all the subtle local, regional and chronological variants of the chipped stone tool-making tradition into a simple portmanteau term, the 'Big Arrowheads Industry' ('BAI' – Kozlowski 1999). A key difference between the projectile points of the early and the later Pre-Pottery Neolithic is the much greater size of the 'big arrowheads'. Behind the detailed shapes of those projectile points there was a widespread use of a new and very particular method of preparing a core and detaching blades from it.

Figure 7.3 A cache of skulls with modelled and painted faces, Tell Aswad. (By courtesy of Danielle Stordeur)

The technique is known as bi-directional or naviform core reduction, meaning that the core was prepared with striking platforms at both ends, so that blades could be removed first from one end of the core, and then from the opposite end. It required great skill and would have been very difficult to learn, as contemporary experimental flint knappers have found (Borrell 2013). At this period, throughout the Levantine corridor and east into the semi-arid region people wanted long blades from which to produce their retouched tools and projectile points. Naviform cores became somewhat boat-shaped because blades were detached alternately from either end of the bipolar core. At that level, there is a common, and highly skilled, reduction procedure that seems to have been developed at the northern end of the Levant, from where it was widely disseminated. It is probable that working with bipolar or naviform core technology was so skilled that it was practised by specialists. It has generally been believed that 'specialisation of labour', that is an economy within which there were specialised craftsmen, scribes, priests, merchants, and soldiers, is the hallmark of urbanised societies and the first civilisations; but here, in the Neolithic, the advanced techniques of stone tool-making imply the preparation and production were the preserve of specialists. There is a second issue: why was such a demanding technique employed in the preparation of beautiful long, parallel-sided blades when the finished product often concealed that expertise? It reminds me of a feature of our contemporary cultures, whereby people can take pride in owning a brand of wristwatch, car, T-shirt or trainers that is generally respected (earning the owner a share in that respect).

At this point, I need to slip in a caveat. There is a danger in identifying this period with the 'agricultural revolution', the phrase which Childe used as a synonym for his 'Neolithic revolution', as if, from this time onwards, people lived by farming. Archaeo-botanists and archaeo-zoologists identified the first recognisably domesticated forms of cereals, and of sheep and goat around the beginning of the late Pre-Pottery Neolithic. Recognising that the cereals or sheep and goat were domesticated means only that they have shown that those plants or those animals have been morphologically changed from their wild forms; domestication in itself tells us nothing about how those plants and animals were used. Although this was covered in Chapters 2 and 3, it deserves to be emphasised here that Neolithic agriculture was what Amy Bogaard and her collaborators call labour-limited as opposed to land-limited agriculture (Bogaard et al. 2018, 2019). Bogaard refers to labour-limited agriculture as 'garden farming', where production is limited by human labour. Cattle were domesticated in different parts of the region only near the end of the Neolithic period, and there is little or no sign that they were used to draw the plough. Farming, as we tend to think of it, with fields ploughed by pairs of oxen, began to appear in southwest Asia beyond the time-frame of this book.

Settlements and Social Organisation

A number of communities of the later Pre-Pottery Neolithic were of considerable size as well as longevity; in fact, there was a wide range of size of settlements, although it is the largest that catch our attention. Within the Mediterranean woodland zone, some settlements – Abu Hureyra, on the Euphrates in north Syria,

'Ain Ghazal in the outskirts of Amman in Jordan, and Basta in southern Jordan for instance – covered 12 ha and more. Çatalhöyük in central Anatolia was more than 13 ha in extent. These settlements at the top end of the range of size housed populations that numbered in thousands; they were the first large-scale communities, with populations that were on a scale of population with many medieval European cities. The range of size of settlements within the southern Levant, where the largest settlements were more than 12 times the size of the smallest, has encouraged some archaeologists to wonder whether there was a settlement hierarchy, with the largest settlements acting as 'central places', the nodes in a nested, hierarchical social and economic network. Mellaart (1967) controversially proposed that Çatalhöyük was a city, and Jane Jacobs (1969) wove Çatalhöyük into her ideas of the origins of urban economies and societies. (Those books were very widely read, and the idea of Çatalhöyük as a city is still commonly encountered.) But there is no evidence of social hierarchy or urban 'central places'; but, on the other hand, nor are they small-scale, simple village communities.

Kenyon made the rectangular mud-brick houses of the later Pre-Pottery Neolithic at Jericho a standard element of her PPNB culture in the southern Levant. But when Diana Kirkbride, who had worked as a member of the Jericho team, went to the south of Jordan to excavate a Neolithic settlement at Beidha, near Petra, she found that the houses were round (Kirkbride 1968). Radiocarbon dating has since shown that they do indeed date within the later Pre-Pottery Neolithic. Only late in the history of the settlement did the inhabitants change to building rectangular houses, but they were not at all like the normal 'pier-houses'; what survived in the excavations were massive basement-level structures sub-divided by stub walls that would have supported a living floor on an upper storey. At Abu Hureyra on the Euphrates in north Syria the extensive later Pre-Pottery Neolithic settlement was made up of rectangular mud-brick houses whose ground floor was made up a number of small storage rooms (Moore et al. 2000). In the tops of the walls there were cylindrical holes that would have held lines of poles that were the main supports of the first-floor walls around the living room floor. Since there were no doorways at ground level, it seems that the living room was accessed by an external ladder, and the ground floor rooms were storage rooms that were accessed from above. One of the characteristics of houses at a number of later Pre-Pottery Neolithic settlements is the presence of substantial and secure storage capacity. And that implies that, in contrast to the communal storage and processing that we saw at early Pre-Pottery Neolithic Jerf el Ahmar, only a few kilometres upstream, the households of Abu Hureyra each took care of their own stores. We are looking at a quite different social organisation of the generally larger communities of the period.

The site of Çayönü Tepesi on a terrace of a small tributary of the upper Tigris, not far from the city of Diyarbakir, has been extensively excavated over many years, allowing us to see something of the pattern of its layout and its very different architecture (Özdoğan 2011). It was first occupied in the early Pre-Pottery Neolithic, when it was a small cluster of circular houses; it grew to its maximum size at the beginning of the later Pre-Pottery Neolithic, and it was finally abandoned around the beginning of the Pottery Neolithic (Figure 7.4).

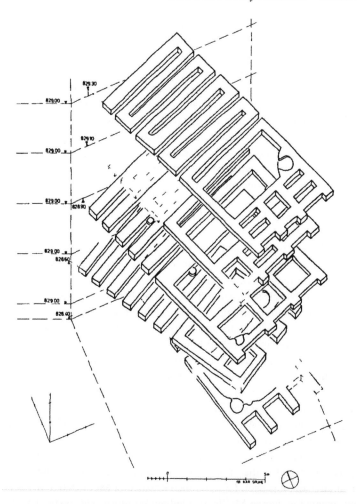

Figure 7.4 Çayönü: the stone foundations of a succession of large houses, each built almost exactly upon its predecessor. While the superstructure was removed completely, the stones from the foundations of the old house were not re-used for the new house. (Schirmer 1990, Fig. 5, with permission)

The history of the settlement has been defined in six phases on the basis of the succession of house designs. After the first phase of roughly circular, semi-subterranean houses, all the houses are large rectangular buildings, averaging 12 m long and 5 m wide; their massive stone foundations have survived (Schirmer 1990). The houses of the second phase were internally sub-divided; half or more of the building had closely spaced transverse walls, presumably sleeper walls supporting a raised wooden living floor. These 'grill-plan houses' were built side by side, all orientated in the same direction, and they were each replaced by building a new layer of foundations on top of the old. The houses of the next phase are known as 'cell-plan'; the interiors were criss-crossed with a grid of walls forming a symmetrical pattern of nine small cells. There were doorways between cells, and there were traces of various activities that were carried out in the cell-like rooms. There was no access to these ground floor cells from the outside, which means that they were accessed from the living floor above, and the living floor was accessed by an outside stair or ladder. The design of the large houses changed again; the living floor was at ground level and consisted of a single room ('large room buildings'), equipped with hearth and oven, and with stone benches against the walls. The settlement seems to have been built around an open central area that was flanked by the largest houses, with lesser houses at some distance from the centre.

The open area at the centre of the settlement had been repeatedly resurfaced with red clay; a row of stone monoliths were set up across it, and to one side there were four 'special buildings' (Özdoğan & Özdoğan 1998). These buildings are unlike the stereotyped domestic buildings; each is quite distinctive and the material they have yielded is different from what is found in the houses. Quite how they relate to each other and to the sequence of architectural house-types is not entirely clear, because they were terraced into the natural subsoil and cannot be related to the stratigraphy of the site. The 'flagstone building' was almost square and consisted of a single semi-subterranean chamber floored with large stone slabs. Close by was the 'skull building', which consisted of an almost square room with stone 'benches' around its walls and a stone-paved floor. The 'skull building' had undergone at least five major rebuilding phases and had been in use for most of the settlement's history. In a pair of subterranean cells in one rear corner it contained deposits of human skulls, skeletons, or skeletal parts. Altogether there were more than 400 individuals represented, together with the horns of wild cattle. In its last form, there was a large flat slab in the main chamber, and when the dark residue on that slab was analysed it was found to be the remains of both human and animal blood (Loy & Wood 1989), tokens of the rituals that were performed there.

There are large and important settlements beyond the arc of the hilly flanks zone. Çatalhöyük, set in the middle of a rich alluvial fan at the southwestern edge of the Konya plain, has been famous since it was first excavated in the early 1960s. The first excavator, James Melllaart, challenged those who were sure that all the meaningful development was to be found in the hilly flanks of the Fertile Crescent, proposing that Çatalhöyük was not only a precocious Neolithic community, but was a proto-urban 'supernova' (Mellaart 1967). His claims were not generally accepted, but the question was how to incorporate such a unique and isolated site and such a large community into the hilly flanks scenario (Figure 7.5).

Ian Hodder has carried out a 25 year research programme at the site, and we now know much more about this determinedly unique community (Hodder 2006, 2014). The site sits in the middle of a rich alluvial fan, and consists of two mounds on either side of one of the fan's braided streams. The eastern mound is the larger and dates to the second half of the later Pre-Pottery Neolithic. It consists of 13 ha of densely packed housing, which us represent a population in the thousands. The rectangular houses were strikingly different from anything in the Levant. They were built of large mud bricks, with the walls or one house abutted against its neighbours; with no access at ground level, houses were generally entered through a trapdoor in the flat roof and a steep ladder. The single-room interiors of the

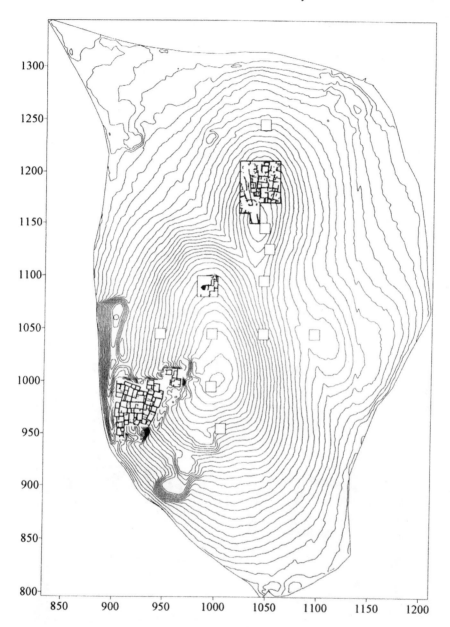

Figure 7.5 Contour plan of Çatalhöyük, the east mound, one of the largest settlements of the later Pre-Pottery Neolithic period. The main areas of excavation are in the southwest of the mound, and the 'north area' near the top of the mound. (By courtesy of Ian Hodder)

houses followed a pattern. The walls and the floors were repeatedly covered with wash after wash of white lime plaster. Some areas of the floor could be coloured red. There was a cooking fire and an oven against the wall immediately below the trapdoor in the roof, and most domestic activities were located on that side of the room (Figure 7.6).

Figure 7.6 Çatalhöyük: plan of the houses in the northern excavation area. (By courtesy of Ian Hodder)

The floor was made up of a series of low platforms, and the platforms opposite the kitchen area were where burials were made under the plaster floor. Some houses had no sub-floor burials, while others had dozens of bodies. Having estimated the average life of a mud-brick house before it had to be taken down and replaced, and having analysed the burials in the search for genetic family relationships, Ian Hodder's research team were clear that the burials did not represent members of the household who had died during the lifetime of the house. Sometimes the walls on that side of the room were given painted decoration, either in the form of hand-prints or patterns, or occasionally with painted scenes. Occasionally an animal's skull was set into the plastered wall; the most extraordinary are the skulls and great spreading horns of several wild aurochs set into a bench-like arrangement in the corner of a living space (Figure 7.7).

While each of the settlements of the Pre-Pottery Neolithic had its own distinct characteristics, they were clearly networked into social and cultural exchange networks with each other. Çatalhöyük, however, seems to have been the only settlement in the area. There were several earlier, smaller settlements, such as Pınarbaşı and Boncuklu (Baird 2012; Baird et al. 2016), but for about a thousand years the whole population of the area was concentrated at Çatalhöyük, whose elaborate belief systems and complex symbolic representations seem to have become more and more intense with time – until things began to change quite radically at the end of the later Pre-Pottery Neolithic (of which more in the next chapter). Over the final centuries of its life, the dense population of earlier times dispersed; the houses were no longer built tight against one another, but were spaced out, and the elaborate wall decorations and intense burial rituals became a thing of the past.

The last centuries of the later Pre-Pottery Neolithic in the Levant saw the rapid formation of a number of very large communities, followed equally quickly by

Figure 7.7 Çatalhöyük: wild bulls' skulls and horns installed in the corner of a room. (By courtesy of Ian Hodder)

the dispersal of their populations The settlement at 'Ain Ghazal, at the edge of the city of Amman, Jordan, is the proto-typical example. The settlement existed from almost the beginning of the later Pre-Pottery Neolithic into the late Neolithic, a total of around 2500 years (Rollefson 2015). It began as a modest village that was perhaps 2 ha in extent, but over the centuries it grew to about 5 ha. In the last part of the later Pre-Pottery Neolithic a number of settlements in Israel and in the Jordan valley were abandoned; but at the same time 'Ain Ghazal expanded rapidly. Like a number of other settlements (they have been collectively called 'mega-sites') it experienced a population explosion, doubling in size within a century and reaching almost 15 ha. As the population expanded, the form of the houses changed; the simple small rectangular houses of the early centuries gave way to massive, multi-roomed, two-storey buildings. Rollefson suggests that simple family households gave way to closely related families pooling their labour and resources into a common store. Then almost as rapidly population declined. Gary Rollefson and his colleagues argued that the growth of such a large population imposed impossible pressures on the local environment on which they depended for cultivation of crops, intensive (and destructive) pasturing of goats, and for timber for building, cooking, and in large quantities for the burning of lime for the making of white lime plaster (Rollefson & Pine 2009).

Ritual, Burials, Curation of Skulls

Once archaeologists had seen the burials and the modelled skulls that were excavated at Jericho, they were alert to undertaking careful excavation and recording when they found burials on their sites. They began to show that many of the burial pits had been re-opened, and that the cranium, or the complete skull and jawbone, had been removed, sometimes involving severing the neck at the cervical vertebrae. At several sites in the southern and central Levant caches of skulls have been found; very occasionally, skulls have been found with facial features modelled in clay, and sometimes also coloured. Some of the skulls have been found to have surfaces that have been polished as if by much handling. Males and females of all ages, including children, can be identified, and we have no idea why the bodies and skulls of a few people were given such special treatment. Ian Kuijt (2008) has teased apart the sequential cycles of ritual that would have accompanied the different stages, reflecting on their role in the social construction of identity and social memory of the community. The burial of the person within the house may have been a small, private, family ritual; at some later time, the burial would have been re-opened for the removal of the cranium or the whole head; skulls were kept for some time, and some of them show signs of having been much handled, perhaps in the context of wider ceremonies that reverenced the ancestors; and finally groups of skulls were given a final resting-place, closing a long cycle of rituals and ceremonies.

The rituals were often particular to an individual community. At most sites, the bodies were buried in an oval grave-pit, lying on one side with knees drawn up tightly to the upper body. But at Tell Hatoula, a settlement beside the Euphrates in north Syria, the burials were concentrated in one building, and the bodies of the dead were bundled into a tightly crouched position and placed sitting upright, knees under chins, in narrow cylindrical pits below the plaster floors of houses (Guerrero et al. 2009). At Tell Aswad, near Damascus in southern Syria, the bodies were placed in a foetal position on the floor against the wall of the house, sometimes within the house and sometimes on the outside. The bodies were then covered with clay and the conspicuous lump was finished with a plaster surface (Stordeur 2003; Stordeur & Khawam 2006, 2007). But there was a further twist in the story, for the custom suddenly changed quite radically. At the edge of the settlement the excavators found two broad shallow pits. Into each of these a clutch of skulls with features carefully modelled in painted plaster had first been placed (Figure 7.3). Then, bodies were given shallow burial in the pits, one after another, apparently in quite rapid succession. Finally, to start the next cycle, almost all of the skeletons had been revisited in order to remove their skulls.

Archaeologists almost always excavate settlements. We are reminded that our knowledge is biased thereby when we consider two extraordinary finds from Israel. The cave of Nahal Hemar in the Judaean desert was discovered when archaeologists were urgently following up the discovery of the Dead Sea Scrolls and the Qumran cave. In the arid conditions there were wooden artefacts, and fragments of rope baskets, woven fabrics and nets. There were also several human skulls with modelled facial features, and strange stone face-masks. Were these the cache of the ritual equipment from some shrine in the valley below the cave? The excavators suggested such a notion (Bar-Yosef & Alon 1988; Goren et al. 1993).

The small site of Kfar HaHoresh is situated in the Galilean hills overlooking a valley where there are several small later Pre-Pottery Neolithic settlements (Goring-Morris 2008; Goring-Morris et al. 1998). At the centre of the site is an area where bodies were buried below limestone plaster surfaces that resemble the plaster floors of houses. To one side was a place where limestone plaster was manufactured and prepared. In another direction was an area of midden, hearths and roasting pits, interpreted as the place where food was prepared, perhaps for ritual feasts. In the central area at least 12 plastered surfaces have been found. The excavator is sure that the surfaces and low walls that edge some of them are not the remains of collapsed buildings but were open-air installations. Many burials have been found beneath these plaster surfaces, but there are also other kinds of burial, notably in stone-built cists. Substantial timber posts or standing stones marked some burials. The remains of more than 50 individuals have been found so far. Some burials are primary interments of single bodies; others are secondary burials or multiple burials. There is evidence that some graves were re-opened for the interment of further bodies. In one grave, two headless adults were buried side by side, one of them with a headless infant in his/her arms and some foetal bones laid in the stomach area. On top of another grave, there was a strange arrangement of bones. Most of the bones were human, but some were gazelle long bones. The bones seem to have been laid out to represent the skeleton of a quadruped, but there is little concern for using the bones in their correct anatomical places. The animal's 'tail' was an articulated human hand, and at least two of its 'hooves' were human mandibles. Since many of the burials were headless, it seems to make sense that there were also a number of burials of separate skulls, either singly or in 'nests'. One cache of four skulls seems to have been plastered, and the excavators suspect that they were buried when the plaster 'faces' were beginning to crumble and detach themselves. It is as if the detached skulls enjoyed a second 'life-cycle' before being buried a second and final time.

A few burials were accompanied by simple grave goods. Chipped stone tools, stone beads, seashells, objects of ground stone and a mother-of-pearl pendant have been found. In a pit a modelled human skull was found with the skeleton of a headless, but otherwise articulated, gazelle. In another burial 250 bones representing portions of at least eight wild cattle accompanied a headless adult human skeleton. Other bodies or skulls were associated with gazelle horn cores, or fox mandibles. Kfar HaHoresh is at present a unique discovery, and we can have no idea whether there are other similar sites awaiting discovery. The features that it presents are in some respects familiar, because similar practices are documented on settlements of the period, and in other respects quite beyond the normal.

The burials within the settlement, and frequently under the floors of houses, the recovery of skulls, and the occasional finds of skulls with modelled facial features have occupied the attentions of excavators, and have produced an extensive literature of discussion of the rituals associated with burial. However, before we leave the subject, we should take note of the fact that no site has produced enough burials to account for the population of the (excavated part of the) settlement. Some sites have produced very few intramural burials, while hundreds of burials have been registered at others, notably Çatalhöyük in central Anatolia; but, while the burials may represent a cross-section of a population from infants to the elderly, they also represent some selective processes. Occasionally, as at 'Ain Ghazal, human remains have been found among the discarded rubbish; by no means all of the people who had lived there received the careful and complex ceremonies that we find fascinating.

Networks of Sharing and Exchange

It is easy for archaeologists to be completely focused on the excavation, recording, and complex analysis of a particular settlement site; and it is equally easy for readers to see the Neolithic in terms of the dots that represent sites on a map. For the people who lived in those settlements it was important to be part of local and regional networks, having social connections with people who lived in other places, exchanging goods and materials, sharing innovations and ideas, and arranging marriage partners. Social exchange systems had operated long before the Neolithic, but the extent and the intensity of networking at local and regional scales grew through the Epipalaeolithic and the Neolithic, changing their nature with increasing sedentism. In earlier times members of a mobile foraging group could rely on meeting others from time to time in the course of their everyday lives; when life in permanent settlements became the norm deliberate efforts were needed in order to maintain the networks of social relations.

The first studies of networking in the Neolithic of southwest Asia were carried out through the medium of obsidian, a black volcanic glass that can be worked like flint (Renfrew & Dixon 1968, 1976). The potential volcanic sources are few, and small numbers of obsidian tools were recorded on almost every Neolithic site. Renfrew and his colleagues, a geologist and a physicist, were able to match the obsidian from early Neolithic sites to one of two sources in central Anatolia or of two more in the east of Turkey, near Lake Van. Renfrew's analysis of the numbers data showed that settlements (like Çatalhöyük) within 200–300 kilometres (125–185 miles) of these locations could supply themselves with the material and use obsidian for their everyday chipped stone tools. Settlements further from the sources, Renfrew hypothesised, relied on contacts from whom they could obtain obsidian in exchange for something else. He found that percentages of obsidian within the chipped stone assemblages at sites declined sharply with distance. At the southern extremes of the distribution – in southern Israel and Jordan, or in southwest Iran, around 900 kilometres (560 miles) from the sources – only one piece of chipped stone in a hundred, or one in a thousand, was obsidian. Clearly, the tiny amounts of obsidian that were obtained in these trades do not represent essential raw materials, but were part of social exchanges. There were other sorts of materials that were also exchanged extensively, such as marine shells from the Mediterranean and the Red Sea, beads of malachite (a vivid green copper oxide) and other greenstone beads, and marble bracelets; but obsidian from archaeological excavations could be sourced and the data could be statistically analysed.

We know that amounts of obsidian in the exchange networks increased over the long term of the ENT: there is much more obsidian at early Pre-Pottery Neolithic sites than at late Epipalaeolithic sites, but there is very much more in circulation in the later Pre-Pottery Neolithic. Recent simulation studies have shown that it is necessary to assume that some people travelled to exchange with partners in societies living some distance away, bypassing their near neighbour communities (Ibañez et al. 2015; Ortega et al. 2014). This is called a small-world or distant link network; while most exchanged with their nearest neighbour communities, a small number had direct links with distant partners, and some settlements became hubs in the increasingly complex and sophisticated networks.

Why was it so important for someone in a community somewhere in the Levant, or in southeast Turkey, or in the valleys of the Zagros Mountains in western Iran, to have a small blade of obsidian, some seashells, or a greenstone bead? In part, possession of such exotic items showed that the owners belonged in the prestigious system of far-flung contacts; at the same time, these special objects of exotic materials ensured the good social relations that were enjoyed between those who had exchanged them with one another. For people to be able to feel that they were part of a large, networked super-community it was necessary to share what they had in common. It has been argued that communities benefited from their investment in these networks because such interaction facilitated the rapid spread of useful knowledge and innovations, and ensured that separate groups recognised each other as partners rather than as strangers and rivals.

Some exchanged items, such as pieces of Anatolian obsidian at settlements in the far south of the Levant, were special because they were exotic. Others, for example the green malachite beads, because they were very rare. From early in the Neolithic period, small tools and beads of worked copper began to be made. Bright, shiny pieces of hot-worked copper were another kind of rare and exotic material that was fed into the extensive social exchange networks.

Cyprus

Cyprus provides this chapter with a coda, as what was happening on the small scale of the island in the later Pre-Pottery Neolithic period presages what will be the central subject of the next chapter. We saw in the previous chapter that there were groups from the mainland who had established settlements, bringing with them not only the traditions of their way of life but also physical resources that the island lacked, such as wheat. In the later Pre-Pottery Neolithic there are substantial settlements all around the island. The best known is Shillourokambos in the village lands of Parekklisha (Guilaine et al. 1995; Guilaine & Vigne 2011). Shillourok-ambos was an open settlement site which was occupied over many centuries. The earliest phase of its occupation is remarkable for its chipped stone industry which is based on the so-called naviform core technique that characterises the beginning of the later Pre-Pottery Neolithic in the Levant. Equally remarkable is the spectrum of animals that provided their meat. It includes sheep, goats and pigs, all of which must originally have been brought from the mainland; and it also includes fallow deer, which, like the wild boar of earlier times, appear to have been introduced wild to provide a prey for hunting. There were other settlements at this period on the north, west and south coasts of Cyprus, as well as in the interior of the island. And the evidence is that these communities were colonists from the mainland who had been at pains to introduce the plants and animals that they needed and that the island lacked. They were preceded by earlier colonists in the late Epipalaeolithic and by more in the early Pre-Pottery Neolithic; but the discovery and excavation of settlements all over the island indicates that the colonisation of the island took off in the later Pre-Pottery Neolithic. And that is surely a signal of population pressure on land and resources in the nearby Levant ahead of the transformation of the settlement map that we will see in the next chapter.

References

Baird, D. (2012). The Late Epipaleolithic, Neolithic, and Chalcolithic of the Anatolian Pla-teau, 13,000–4000 BC. In D. T. Potts (Ed.), *A Companion to the Archaeology of the An-cient Near East* (pp. 431–66). Oxford: Wiley-Blackwell.

Baird, D., Fairbairn, A., & Mustafaoğlu, G. (2016). Boncuklu: The Spread of Farming and the Antecedents of Çatalhöyük. *Heritage Turkey*, 6, 15–8. doi:10.18866/biaa2016.027

Bar-Yosef, O., & Alon, D. (1988). Nahal Hemar Cave. *'Atiqot. English series*, 18, 1–30.

Bogaard, A., Fochesato, M., & Bowles, S. (2019). The Farming-Inequality Nexus: New Insights from Ancient Western Eurasia. *Antiquity*, 93(371), 1129–43. doi:10.15184/aqy.2019.105

Bogaard, A., Styring, A., Whitlam, J., Fochesato, M., & Bowles, S. (2018). Farming, In-equality, and Urbanization: A Comparative Analysis of Late Prehistoric Northern Meso-potamia and Southwestern Germany. In T. Kohler & M. Smith (Eds.), *Ten Thousand Years of Inequality: The Archaeology of Wealth Differences* (pp. 201–29). Tucson: University of Arizona Press.

Borrell, F. (2013). Opening Pandora's Box: Some Reflections on the Spatial and Temporal Distribution of the Off-Set Bi-Directional Blade Production Strategy and the Neolithisation of the Northern Levant. In F. Borrell, J. J. Ibáñez, & M. Molist (Eds.), *Stone Tools in Transition: From Hunter-Gatherers to Farming Societies in the Near East* (pp. 247–64). Barcelona: Universitat Autònoma de Barcelona, Servei de Publicacions.

Byrd, B. F., & Banning, E. B. (1988). Southern Levant Pier-Houses: Intersite Architectural Patterning during the Pre-Pottery Neolithic B. *Paléorient*. 14(1), 65–72.

Goren, Y., Segal, I., & Bar-Yosef, O. (1993). Plaster Artifacts and the Interpretation of the Nahal Hemar Cave. *Journal of the Israeli Prehistoric Society*, 25, 120–31.

Goring-Morris, A. N. (2008). Kfar Ha-Horesh. In E. Stern (Ed.), *The New Encyclopedia of Archaeological Excavations in the Holy Land* (pp. 1907–9). Jerusalem; Washington, DC: Israel Exploration Society and Biblical Archaeology Society.

Goring-Morris, A. N., Burns, R., Davidzon, A., Eshed, V., Goren, Y., Hershkovitz, I., . . . Kelecevic, J. (1998). The 1997 Season of Excavations at the Mortuary Site of Kfar HaHoresh, Galilee, Israel. *Neo-Lithics: A Newsletter of South-West Asian Lithics Research*, 3/98, 1–4.

Guerrero, E., Molist, M., Kuijt, I., & Anfruns, J. (2009). Seated Memory: New Insights into Near Eastern Neolithic Mortuary Variability from Tell Halula, Syria. *Current Anthropology*, 50(3), 379–91. doi:10.1086/598211

Guilaine, J., Coularou, J., Briois, F., & Carrère, I. (1995). 1994 Work Carried on in Collaboration with the Ecole Française d'Athènes: The Neolithic Site of Shillourokambos. *Bulletin De Correspondance Hellenique*, 119(2), 737–41.

Guilaine, J., & Vigne, J. D. (2011). *Shillourokambos: un établissement néolithique précéramique à Chypre: les fouilles du secteur 1*. Paris: Errance–École Française d'Athènes.

Hodder, I. (2006). *The Leopard's Tale: Revealing the Mysteries of Turkey's Ancient 'Town'*. London: Thames & Hudson.

Hodder, I. (2014). Çatalhöyük: The Leopard Changes Its Spots. A Summary of Recent Work. *Anatolian Studies*, 64, 1–22. doi:10.1017/S0066154614000027

Ibañez, J. J., Ortega, D., Campos, D., Khalidi, L., & Méndez, V. (2015). Testing Complex Networks of Interaction at the Onset of the Near Eastern Neolithic Using Modelling of Obsidian Exchange. *Journal of The Royal Society Interface*, 12, 20150210. doi:10.1098/rsif.2015.0210

Jacobs, J. (1969). *The Economy of Cities*. New York, NY: Random House.

Kenyon, K. (1957). *Digging up Jericho*. London: Ernest Benn.

Kenyon, K. M. (1960). *Archaeology in the Holy Land* (4th ed.). London: Benn.

Kenyon, K. M., & Holland, T. A. (1981). *Excavations at Jericho, Vol. 3, The Architecture and Stratigraphy of the Tell*. London: British School of Archaeology in Jerusalem.

Kenyon, K. M., & Holland, T. A. (1982). *Excavations at Jericho. Vol. 4, The Pottery Type Series and Other Finds*. London: British School of Archaeology in Jerusalem.

Kenyon, K. M., & Holland, T. A. (1983). *Excavations at Jericho, Vol. 5: The Pottery Phases of the Tell and Other Finds*. Oxford: British School of Archaeology in Jerusalem & Oxford University Press.

Kirkbride, D. (1968). Beidha: Early Neolithic Village Life South of the Dead Sea *Palestine Exploration Quarterly*, 42(168), 263–74.

Kozlowski, S. K. (1999). The Big Arrowhead Industries (BAI) in the Near East. *Neo-Lithics*, 2/99, 8–10.

Kuijt, I. (2008). The Regeneration of Life: Neolithic Structures of Symbolic Remembering and Forgetting. *Current Anthropology*, 49(2), 171–97.

Loy, T. H., & Wood, A. R. (1989). Blood Residue Analysis at Çayönü Tepesi, Turkey. *Journal of Field Archaeology*, 16(4), 451–60.

Mellaart, J. (1967). *Çatal Hüyük: A Neolithic Town in Anatolia.* London: Thames and Hudson.

Moore, A. M. T., Hillman, G. C., & Legge, A. J. (2000). *Village on the Euphrates: From Foraging to Farming at Abu Hureyra.* Oxford: Oxford University Press.

Ortega, D., Ibañez, J., Khalidi, L., Méndez, V., Campos, D., & Teira, L. (2014). Towards a Multi-Agent-Based Modelling of Obsidian Exchange in the Neolithic Near East. *Journal of Archaeological Method and Theory*, 21(2), 461–85. doi:10.1007/s10816-013-9196-1

Özdoğan, A. (2011). Çayönü. In M. Özdoğan, N. Başgelen, & P. Kuniholm (Eds.), *The Neolithic in Turkey. New Excavations and New Research - The Tigris Basin* (pp. 185–269). Istanbul: Arkeoloji ve Sanat Yayinlari.

Özdoğan, M., & Özdoğan, A. (1998). Buildings of Cult and the Cult of Buildings. In M. Mellink, G. Arsebük, & W. Schirmer (Eds.), *Light on Top of the Black Hill, Studies Presented to Halet Çambel* (pp. 581–93). Istanbul: Ege Yayınları.

Renfrew, C., & Dixon, J. E. (1968). Further Analyses of Near Eastern Obsidians. *Proceedings of the Prehistoric Society*, 34, 319–31. doi:10.1017/S0079497X0001392X

Renfrew, C., & Dixon, J. E. (1976). Obsidian in West Asia: A Review. In G. Sieveking (Ed.), *Problems in Economic and Social Archaeology* (pp. 137–50). London: Duckworth.

Rollefson, G. O. (2015). 'Ain Ghazal (Jordan): A Neolithic Town Prima Inter Pares. In G. Barker & C. Goucher (Eds.), *The Cambridge World History, Volume 2: A World with Agriculture, 12,000 BCE–500 CE* (pp. 243–60). Cambrdige: Cambridge University Press.

Rollefson, G., & Pine, K. J. (2009). Measuring the Impact of LPPNB Immigration into Highland Jordan. *Studies in the History and Archaeology of Jordan*, 10, 473–81.

Schirmer, W. (1990). Some aspects of the building in the "aceramic neolithic" settlement at Çayönü Tepesi. *World Archaeology*, 21(3), 363–87.

Stordeur, D. (2003). Des crânes surmodelés à Tell Aswad de Damascène. (PPNB - Syrie). *Paléorient*, 29(2), 109–16.

Stordeur, D., & Khawam, R. (2006). L'aire funéraire de Tell Aswad (PPNB). *Syria*, 83, 39–62.

Stordeur, D., & Khawam, R. (2007). Les crânes surmodelés de Tell Aswad (PPNB, Syrie): premier regard sur l'ensemble, premières réflexions. *Syria*, 84, 5–32.

8 Further Transformation

Dispersal and Expansion

In the late Pre-Pottery Neolithic, we saw that large, sedentary communities flourished, some of them with populations of many hundreds or even a few thousand people. Simple garden-farming had become established. Around 6500 BCE, things began to change in ways that deserve to be recognised as a further transformation. There were changes all around the arc of the hilly flanks and also in central Anatolia. This transformed late Neolithic period had the greatest implications for the future developments both within the curve of that arc, and for the outward spreading of the Neolithic all around the periphery of what has been the core zone of this book.

As far as archaeologists are concerned, the most notable component of the transformation was the appearance of pottery. Pottery survives well; it broke easily, so there is usually plenty of it, and the clay, the shapes, the firing, and the surface decoration give the archaeologist plenty to work with. Since Kathleen Kenyon defined the Pre-Pottery Neolithic in the deep stratigraphy of Tell es-Sultan, ancient Jericho, the overlying stratum was labelled Pottery Neolithic (or PN). The Pottery Neolithic stratum at Jericho was much more difficult to make sense of, probably because the mud-brick buildings were less substantial and more spaced out, leading to rapid erosion when any of them were abandoned. It was more difficult to find good samples for radiocarbon dating, which at that time required large lumps of carbonised wood. And relating the Jericho PN cultural assemblage to what was already known at other sites in the Levant was far from simple because the PN did not present a single, homogeneous form. Yosef Garfinkel has described the PN period in the Levant as 'the creation of scientific chaos' (Garfinkel 2020: 1439).

Fewer archaeologists have been interested in the PN period; the later Pre-Pottery Neolithic has been considered as 'the culmination of the lengthy shift to productive economies' (Goring-Morris & Belfer-Cohen 2020: 3), so research has continued to be focused on the exploration of the further questions that the Pre-Pottery Neolithic had thrown up. Interest in the rest of the Neolithic has flagged; but, as we shall see, the story is far from over. The evidence shows that the new way of life continued to be successful; indeed, it can be argued that the PN way of life was, in evolutionary terms, more successful, and, in historical terms, more significant.

All around southwest Asia this final phase in the Neolithic leads into the following periods, when crafts such as pottery and copper metallurgy began to become

DOI: 10.4324/9781351069281-9

specialised industries, and when proto-urban centres grew to become by the late fourth millennium BCE the first city-states. In that sense, the PN period is the foundation on which Gordon Childe's urban revolution is built. And all around the peripheries of southwest Asia, new Neolithic societies appeared and expanded. The rest of this chapter looks briefly sat the various regions and their different responses in turn.

The Levant

The climax of the Pre-Pottery Neolithic was short-lived. It was noted a long time ago that aceramic Neolithic settlements in the Mediterranean woodland zone west of the Jordan valley began to be abandoned, leaving a hiatus of several centuries before late Neolithic pottery-using settlements appear. In the highlands east of the Jordan valley, a string of settlements ('Ain Ghazal is the best documented example) (Rollefson et al. 2014, 2020), Motza, near Jerusalem, is the latest to be added to the list (Khalaily et al. 2020) grew rapidly as they accumulated people who had left settlements further west. They exceed 10 ha in extent, and have been called 'mega-sites', their populations growing rapidly to perhaps several thousand people each in the latter part of the later Pre-Pottery Neolithic. But the mega-sites were short-lived: by 6500 BC or soon after, their populations dispersed and they were reduced to small villages of scattered houses, or were abandoned. Further north, Abu Hureyra beside the Euphrates in north Syria, which had grown to be a large farming community in the late Pre-Pottery Neolithic, declined to a scattering of simple houses (Moore et al. 2000). The settlement failed as the Pottery Neolithic was beginning, and the site was abandoned. The same is true of Çayönü in south-east Turkey, which had been in existence for almost two millennia (Özdoğan 2011).

The implosion of population at the south Levantine mega-sites corresponds to the appearance of small settlements dispersed across a much wider range of landscapes. In the Mediterranean woodland zone, villagers depended on mixed farming; east of the rift valley and the Dead Sea, for example at 'Ain Ghazal, small communities consisted of some families who practised some agriculture combined with the herding of sheep and goats, while other families spent part of the year in the village, but most of the year reliant on transhumant pastoralism. In the semi-arid of northern Jordan Andrew Garrard and his teams working in the Wadi Jilat explored sites that date to precisely the last centuries of the Pre-Pottery Neolithic and the overlapping earliest Pottery Neolithic in the seventh millennium BCE (Garrard et al. 1994). These small communities relied predominantly on the herding of their flocks of domesticated sheep and goat.

For a long time there has been debate about whether these small communities in the semi-arid interior had come from the classic late Pre-Pottery Neolithic settlements of the Mediterranean woodland on either side of the Jordan valley), bringing with them their domesticated animals (as argued, for example, by Ilse Köhler-Rollefson (1992) on the basis of her study of the 'Ain Ghazal material), or whether they were the descendants of indigenous hunter-gatherer groups who had obtained domesticated sheep and goat that enabled them to become nomadic pastoralists (as proposed by Douglas Baird (1994)). We now have a neat study that has carried out isotopic analysis of the $\delta^{13}C$ and $\delta^{15}N$ in the collagen of sheep and goat bones from two of the Wadi Jilat sites and from 'Ain Ghazal; the levels of those isotopes inform on the diet of the animals, reflecting the difference between a diet based on plants like grasses and cereals that require temperate conditions and plants characteristic of arid environments (Miller et al. 2019). The results from the animal bones of one of the Wadi Jilat sites, combined with a detailed study of the lithic tradition, suggest an indigenous steppic community that was undertaking animal herding as a continuation of their mobile strategies. The results from the other Wadi Jilat site were more nuanced; it seems that this small indigenous community obtained some of their livestock, and some innovations in chipped stone tool-making, from contacts in the greener landscapes at some distance to the west. The isotopic analyses of the animals from 'Ain Ghazal showed that they had not grazed in the semi-arid interior at a distance from the settlement, suggesting that seasonal nomadic pastoralism was not practised there. The two Wadi Jilat sites are only a few hundred metres distant from one another, suggesting that they may have been occupied at different times. The differences in their pastoralism strategies tells us that each small community made its own way of life, one of them completely self-reliant, and the other perhaps obtaining some animals that were bred in the greener region to the west, and at the same time also picking up some new tool-making tricks.

Beyond the Azraq area, in the far northeast of Jordan, the extensive basalt 'Black Desert' seems an unlikely region to find Neolithic people. There have been many signs of human activity, including 'desert kites' (hunting traps), cairns, extensive wandering walls, and strange wheel-like enclosures, but accessing them on the ground and carrying out investigations is very challenging. Pioneer work on the wheel-shaped enclosures by Alison Betts (Betts 1985; Betts et al. 1998) has been followed by the first investigations that have produced dates for one of the wheel-enclosures (Rollefson et al. 2016; Rowan et al. 2015). There is now good evidence that at least some of the desert kites and the wheel-enclosures began to appear in the late Neolithic. These seem to be new kinds of sites used by groups who colonised the deep arid interior, focusing on places where the meagre rainfall produced seasonal green spots that could support very limited cultivation combined with pastoralism (Rollefson 2011; Rollefson et al. 2014).

Mesopotamia

Coinciding with the time when sites, like Çayönü, in the piedmont and the upper Tigris and Euphrates drainages in southeast Turkey were abandoned, a rash of small settlements sprang up across north Syria east of the Euphrates and north Iraq both east and west of the Tigris. They were situated in the narrow strip of green land south of the hills and north of the semi-arid steppe interior, the Jezirah. The first to be investigated was Tell Hassuna, in north Iraq (Lloyd et al. 1945). The settlement had been long-lived, and at the base of the mound, below layers that produced painted pottery of kinds that were already known from Chalcolithic period sites, the excavators found a new kind of ceramic assemblage. There were simple, well-made jars and bowls, some plain, some with incised decoration, and some painted in red on a cream background. In the 1970s a team from the USSR found a number of prehistoric sites near Telafar, west of Mosul (Merpert & Munchaev 1987; summarised by Yoffee & Clark 1993). They excavated what we can now recognise as a late Pre-Pottery Neolithic site, Maghzaliyah, beside a stream on the southern slopes of the Jebel Sinjar hills; and they also explored a cluster of settlement mounds, Yarim Tepe I–IV, out on the plain at the foot of the hills. Yarim Tepe I produced pottery very like Tell Hassuna, but the lowest levels of the site showed that that ceramic tradition had existed in an earlier form, which became known as Proto-Hassuna. From their work we can see that the last Pre-Pottery Neolithic occupation of the slope of the hill country, from which a range of different ecological zones could be exploited, was succeeded by the colonisation of the open plain, which implies a full commitment to a mixed agricultural economy.

When the late Tony Wilkinson led a survey team on the North Jezirah survey project in the 1980s, covering the plain north of the Jebel Sinjar hills, the picture became clearer (Wilkinson & Tucker 1995). The survey area was a large triangle north of the Jebel Sinjar hills, east of the Iraqi-Syrian frontier, and west of the Tigris valley. The recording of all the prehistoric sites with systematic surface collection and analysis of pottery showed that there was a steady increase in the number and size of settlements, starting with the Proto-Hassuna in the last centuries of the seventh millennium BCE and continuing through the Hassuna and on through the Chalcolithic towards the emergence of urban landscapes in the Bronze Age. The transfer from sites like the late Pre-Pottery Neolithic Maghzaliyah to the Pottery Neolithic villages of the plain was clearly successful (Figure 8.1).

We should not think of the occupation of the Jezirah zone as consisting of uniformly small villages. Peter Akkermans led a team from Leiden University that excavated a cluster of low mounds, Tell Sabi Abyad I-IV, beside the Balikh river in northeast Syria (Akkermans 2013). The Balikh collects its waters over an extensive area in southeast Turkey and flows south to join the Euphrates near Deir ez-Zohr. The Balikh headwaters area is full of PN and Chalcolithic sites, but Tell Sabi Abyad is further south, where the Balikh has become a single stream flowing through a landscape that very rapidly turns from green to semi-arid steppe. The cluster of settlements was inhabited between about 6500 and 5800 BCE. One (burnt) layer at the largest of the mounds, dating to the last two centuries of the seventh millennium, has been intensively excavated and meticulously studied. This settlement consisted of extensive, closely spaced rectangular granaries and storehouses surrounded by a number of small, round dwellings, which all ended in a conflagration around 6000 BCE. Rich inventories have been recovered from the burnt buildings, including several hundred clay sealings. These lumps of clay had been applied on the knots that fastened covers on ceramic storage jars, baskets, and stone vessels; they had then been impressed with small, circular seals with various simple designs (Akkermans & Duistermaat 1997, 2014). Akkermans and his research team concluded that the population of the settlement a large number of pastoralists who regarded the settlement as their permanent home, but who spent parts of the year away from the site with their flocks of sheep and goat. The pottery from

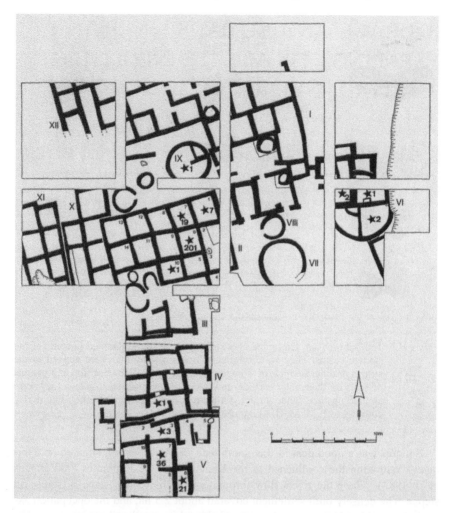

Figure 8.1 Tell Sabi Abyad, plan of the "burnt village". Most of the ranges of rooms are storage facilities, and there are few houses. (Akkermans & Duistermaat 1997, Fig. 1. By courtesy of P. M. M. G. Akkermans)

this settlement provided samples whose analysis showed that some had contained milk products while others had been used for stewing meat (Evershed et al. 2008; Nieuwenhuyse et al. 2015). The important point is that people were beginning to use milk (making yoghurt and cheese) in addition to keeping animals for meat. Long before it became possible to identify traces of lipids in the surface of pottery, Andrew Sherratt (1981) first hypothesised this kind of diversification when he proposed a 'secondary products revolution'. Akkermans (2023) has also noted that spindlewhorls made their first appearance at this time, suggesting that the spinning of woollen yarns and the weaving of textiles were further examples of economic innovation and diversification (Figure 8.2).

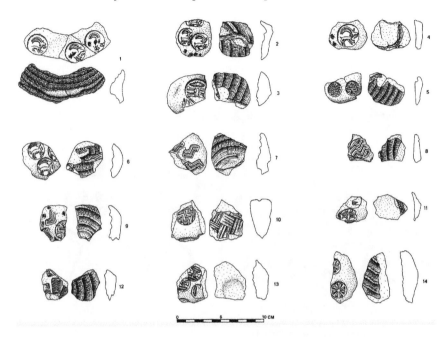

Figure 8.2 Tell Sabi Abyad. Discarded clay sealings were found dumped in some of the
storage rooms. They were originally lumps of clay that were applied to the
strings or other sealings of storage vessels or sacks. They bear the impressions
of seals on their outer surface, and traces of the material to which they were
applied on their inner surface. (Akkermans & Duistermaat 1997, Fig. 4. By
courtesy of P. M. M. G. Akkermans)

Bouqras was a good deal further south than Tell Sabi Abyad, situated on a ter- .
race overlooking the confluence of the Khabur with the Euphrates (Akkermans
et al. 1983), where the rivers flow through an extensive landscape of semi-arid
steppe. The densely built settlement covered about 2.5 ha, and the occupation extended over about 500 years in the second half of the seventh millennium BCE. The
repeatedly rebuilt mud-brick buildings were densely packed and had built up a stratigraphy of more than 5 m. It must have been a substantial community of several
hundred people. Living in a region with only around 125 mm of annual rainfall,
they cultivated wheat, barley, lentils and peas on the narrow floodplain of the Euphrates, but they must have relied heavily on their flocks of sheep and goat. Their
pottery was simple and some of the earliest known. They also used some beautiful
stone vases and vessels made out of gypsum (called by archaeologists whiteware).
More importantly, the site shows that for half a millennium a substantial population
made a successful and sedentary way of life that relied heavily on pastoralism in a
rather unlikely landscape (Figure 8.3).

Another settlement that deserves mention was Choga Mami, situated north-east
of Baghdad, and dating in the sixth millennium BCE, somewhat later than Bouqras.
It was a small settlement whose importance is that evidence of the existence of an

Figure 8.3 Seal impressions on clay sealings. Tell Sabi Abyad. (By courtesy of P. M. M. G. Akkermans)

irrigation channel was trapped in the stratigraphy at the edge of the mound (Oates 1969). Here was another community that had found a way to make a productive economy in an environment that lacked sufficient natural rainfall, presaging the extensive irrigation-landscapes that supported later urban populations. In the heart of the delta Tigris-Euphrates alluvium south of Baghdad it has been difficult to locate and investigate early settlements, which can be deeply buried in silts and below the water table. An exception is the site of Tell Oueili, a settlement where the French excavators were able to penetrate through 4.5 m of archaeological deposit before reaching the water table (Huot 1992, 1996). The lowest accessible levels took the story of settlement in the alluvial south of Mesopotamia further back than any previous site (to around 6000 BCE). The settlement was made up of substantial mud-brick houses with a distinctive tripartite plan, whose central element was a large pillared hall. Traces of cereal crops, mainly barley, indicate that some form of irrigation must have been practised, and there were also date palms. The spectrum of farmed animals consisted of rather few sheep and goat; about 38% of the animal bones were of pig, and 45.5% were cattle (Figure 8.4).

One more Pottery Neolithic site in Iraq is important to mention. Umm Dabaghiyah was excavated in the early 1970s, and it is unlike any settlement site of the early Pottery Neolithic period. It is situated deep in the semi-arid steppe southwest of Mosul and far to the west of the Tigris. There are obvious questions of why there was a settlement in such an unlikely location. It is a very strange place, and certainly not a normal settlement: there were only three of four small, simple mud-brick houses, and the main structures were two long buildings

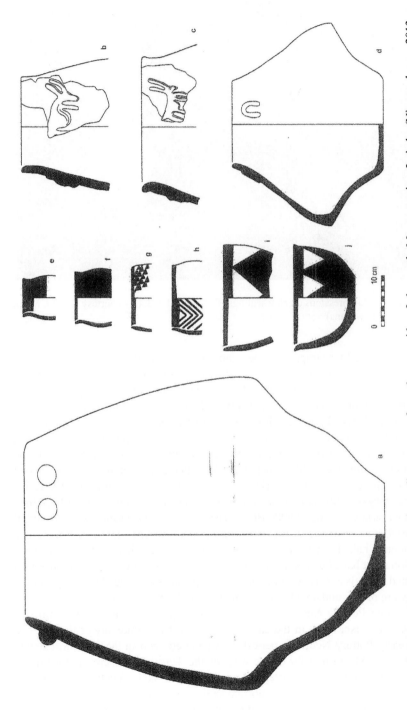

Figure 8.4 Proto-Hassuna pottery, the earliest pottery from the east side of the north Mesopotamian Jezirah. (Nieuwenhuyse 2013, Fig. 3. With permission of ex oriente)

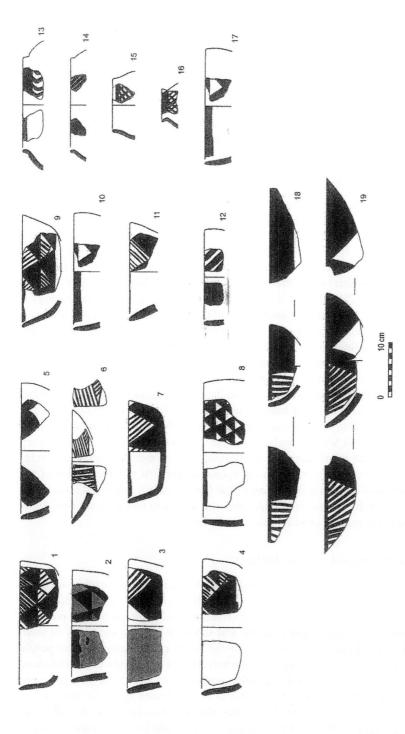

Figure 8.5 The pottery of the Pottery Neolithic rapidly became more sophisticated and technically advanced. (Nieuwenhuyse 2013, Fig. 6. With permission of ex oriente)

parallel to one another, each consisting of two rows of small cells, each about 1.6 m². There were sparse signs of the presence of some cultivated crops, but very few sheep and goat bones. There were large quantities of bones of onager (68.5%) and gazelle (16%), which the excavator, Diana Kirkbride, noted both produce fine quality hides. She therefore suggested that the settlement was 'a trading outpost established for the express purpose of collecting fine hides' (Kirkbride 1974). It is hard to disagree with her reasoning, but it leads to two further, unanswerable, questions: who was hunting those animals in the steppe-desert, and, more importantly, who was prizing the fine animal hides, and how did they use them? The implication of Umm Dabaghiyah is that it fed specialist products into a diverse economy that was not a simple, two-dimensional society of self-sufficient farming villages (Figure 8.5).

The Deh Luran plain in southwest Iran is adjacent to the eastern edge of the Tigris-Euphrates alluvium in southern Iraq. Frank Hole and Kent Flannery had worked there in the early 1960s, excavating the site of Ali Kosh (Hole et al. 1969; Hole & Flannery 1968). The plain is where several small rivers coming out of the Zagros mountains run out in marshy hollows in the alluvium. There had been Neolithic communities living on the alluvium of the Deh Luran plain in southwest Iran from at least 7500 BCE. They had arrived from the Zagros valleys looking for places to settle. Ali Kosh was established alongside a patch of seasonal wetland; the settlement flourished and grew as their cultivation of cereals flourished, gradually supplanting their hunting and gathering. The sheep and goats that they had brought with them became substantial flocks of domesticated animals. With time the number of settlements also multiplied.

We have now traversed the arc around the inner edge of the hilly flanks zone, where new and very different kinds of settlement and new versions of agriculture and pastoralism flourished in environments very different from those of the hilly flanks. Across north Mesopotamia, and in the southern alluvium the late or Pottery Neolithic laid the foundations for the development of more complex economies and societies, leading towards the emergence of the first urban societies of the fourth and third millennia BCE. East of the Mediterranean woodland zone of the Levant, where the landscapes rapidly became semi-arid and arid, we now know that there developed successful pastoralist communities.

Anatolia, the Aegean, and the Balkans

At Çatalhöyük in central Turkey, around 6500 BC, the intense occupation of the site began to change rapidly. Many of the houses that were in use around 6600–6500 BCE were among the most elaborately equipped with painted plaster, plastered animal skulls and wild aurochs horns, and sub-floor burials; and many of them ended their use-lives in what seem to be deliberate conflagrations. Many people left and set up smaller settlements elsewhere around the Konya plain, and those that remained lived in spaced out, independent houses with large storage-spaces and much less attention to sub-floor burial and elaborate symbolic installations (Baird 2012; Marciniak 2015). As late Neolithic settlements spread around Çatalhöyük, new settlements also began to appear in the southwest of Anatolia, and, a little later, in northwest Anatolia, into European Turkey and the Balkans. A few years ago it was suggested that the spread of Neolithic populations beyond the core area of the hilly flanks and central Anatolia was in consequence of a rapid climatic change event that had been detected at the end of the seventh millennium BCE (B. Weninger et al. 2009). However, the earliest settlements on the west coast of Turkey and on the Aegean islands closest to that coast have now been shown to have been in existence by at least 6700 BCE. This suggests that there was a maritime expansion of colonists originating from the northeast corner of the Mediterranean that arrived in the Aegean before the land-based westward expansion of farming settlements reached the coast (Çilingiroğlu et al. 2012; Horejs et al. 2015). The question of whether the source of this westward expansion of Neolithic population originated in central Anatolia or at the north end of the Levant is debated, and at present the evidence for either hypothesis is ambivalent. These new communities on the Aegean had access to obsidian from the island of Melos, but they also obtained a small amount from Göllüdağ in Cappadocia. Brami (2017) has sought to show that the westward wave of colonists starting from the north of the Levant and central Anatolia carried with them a specific idea of what a house should be. They enjoyed relations with mainland Anatolian communities, but they were also able to find their place within the (as yet poorly documented) pre-existing maritime network.

As was the case much earlier with Cyprus, the first colonists on the Aegean coast of Turkey were joined by successive waves of more colonists. At the end of the seventh millennium BCE the houses at Ulucak become crowded together, and the simple burnished pottery of the earliest settlers is joined by pottery known from Cilicia and the Hatay of coastal southeast Turkey, its surface decorated with impressions made with the edges of shells or other small tools (Çilingiroğlu et al. 2012). From this stage on, the expansion of Neolithic communities 'takes off' at an increasingly rapid rate: colonising groups, recognisable by their impressed ware ceramics, appear in Italy and Sicily, on the Riviera coast of southern France, and in Spain and the Balearics. On the basis of the radiocarbon dates, the spread of Neolithic communities across the Balkans and central Europe also took place at an extraordinary rate: the spread of the Neolithic from the Aegean to northeast Hungary is estimated to be more than 1000 km in less than two centuries, or around 5 km per year (Weninger et al. 2014: 2). We need to remind ourselves that this spread, which has to be matched by a similar, but as yet not well documented, spread within southwest Asia, and to the north, south and east. It represents the spread of a rapidly increasing population. Within the Fertile Crescent, there are some areas that were populous in the later Pre-Pottery Neolithic, but seem to be almost abandoned in the Pottery Neolithic; but these areas are few and small, and generally the new sites in the Levant, across the Jezirah in north Syria and Iraq, and on the alluvium in southern Mesopotamia and southwest Iran must represent expanding population numbers.

Conclusion

This is not the place to look back across the whole transformation process from the early Epipalaeolithic through to the late Neolithic of this chapter; that will follow in Chapter 10. Here, I want only to make the point that the transformation was not complete with the climax of the later Pre-Pottery Neolithic. Since Kathleen Kenyon was discovering the Pre-Pottery Neolithic at Jericho, and Robert Braidwood was initiating research into the origins of agriculture in the 1950s, research into the domestication of plants and animals has focused attention on the Pre-Pottery Neolithic, to the detriment of the late Neolithic. Also, the Pre-Pottery Neolithic settlements proved such solid masses of buildings and stratigraphy, while late Neolithic settlements were presented more fugitive remains.

But this complicated chapter shows that there were further transformations at the end of the Pre-Pottery Neolithic and through the Pottery Neolithic periods. Most significantly, we see that population continued to expand. At a certain stage in the later Pre-Pottery Neolithic we see that people began to choose to live in communities that were smaller than the average later Pre-Pottery Neolithic settlements, and certainly much smaller than the so-called mega-sites. But population levels in general continued to rise, and the spreading of late Neolithic populations in southwest Asia were fed on new versions of the Neolithic economy, such as transhumant pastoralism, and irrigation-fed agriculture. We see new technologies, such as the early experiments with metal-working, but most obviously in the rapid development of ceramics, as containers, as cooking and food preparation pots, and as skilfully made and decorated 'table-ware'. And the new kinds of ceramic relate to expansions in diet as well as manners of presenting food. Theya Molleson, who studied the human remains from Abu Hureyra, was able to observe the changing dietary customs in their teeth (Molleson 1994, 2000). There was much that was new in the late Neolithic; and it is the period that saw the beginnings of a rapid outward expansion of Neolithic populations in all directions. It was, in several ways, a period of transformation in its own right.

References

Akkermans, P. A., Boerma, J. A. K., Clason, A. T., Hill, S. G., Lohof, E., Meiklejohn, C., ... van Zeist, W. (1983). Bouqras Revisited: Preliminary Report on a Project in Eastern Syria. *Proceedings of the Prehistoric Society*, 39, 335–72. doi:10.1017/S0079497X00008045

Akkermans, P. M. M. G. (2013). Tell Sabi Abyad, or the Ruins of the White Boy: A Short History of Research into the Late Neolithic of Northern Syria. In D. Bonatz & L. Martin (Eds.), *100 Jahre archäologische Feldforschungen in Nordost-Syrien – eine Bilanz* (pp. 29–43). Wiesbaden: Harrassowitz Verlag.

Akkermans, P. M. M. G. (2023). Earliest Date for Seals and Sealings in the Near East. In L. E. Bennison-Chapman (Ed.), *Bookkeeping without Writing: Early Administrative Technologies in Context* (pp. 35–55). Leiden; Leuven: Nederlands Instituut voor het Nabije Oosten & Peeters.

Akkermans, P. M. M. G., & Duistermaat, K. (1997). Of Storage and Nomads - The Sealings from Late Neolithic Sabi Abyad, Syria. *Paléorient*, 22, 17–44.

Akkermans, P. M. M. G., & Duistermaat, K. (2014). The Late Neolithic Seals and Sealings. In P. M. M. G. Akkermans, M. L. Bruning, H. O. Huigens, & O. P. Nieuwenhuyse (Eds.), *Excavations at Late Neolithic Tell Sabi Abyad, Syria. The 1994–1999 Field Seasons* (pp. 113–23). Turnhout: Brepols.

Baird, D. (1994). Chipped stone production technology from the Azraq project Neolithic sites. In H.-G. Gebel & S. K. Kozlowski (Eds.), *Neolithic Chipped Stone Industries of the Fertile Crescent* (pp. 525–42). Berlin: Ex Oriente.

Baird, D. (2012). The Late Epipaleolithic, Neolithic, and Chalcolithic of the Anatolian Plateau, 13,000–4000 BC. In D. T. Potts (Ed.), *A Companion to the Archaeology of the Ancient Near East* (pp. 431–66). Oxford: Wiley-Blackwell.

Betts, A. V. G. (1985). Black Desert Survey, Jordan: Third Preliminary Report. *Levant*, 17, 29–52.

Betts, A., Colledge, S., Martin, L., McCartney, C., Wright, K., & Yagodin, V. (1998). *The Harra and the Hamad.* Sheffield: Sheffield Academic.

Brami, M. (2017). *The Diffusion of Neolithic Practices from Anatolia to Europe. A Contextual Study of Residental Construction, 8500–5500 BC cal.* Oxford: BAR International Series 2838.

Çilingiroğlu, A., Çevik, Ö., & Çilingiroğlu, Ç. (2012). Ulucak Höyük. Towards Understanding the Early Farming Communities of Middle West Anatolia: The Contribution of Ulucak. In M. Özdoğan, N. Başgelen, & P. Kuniholm (Eds.), *The Neolithic in Turkey. New Excavations & New Research. Volume 4. Western Turkey* (pp. 139–75). Istanbul: Archaeology & Arts Publications.

Evershed, R. P., Payne, S., Sherratt, A. G., Copley, M. S., Coolidge, J., Urem-Kotsu, D., . . . Burton, M. M. (2008). Earliest Date for Milk Use in the Near East and Southeastern Europe Linked to Cattle Herding. *Nature*, 455(7212), 528–31. doi:10.1038/nature07180

Garfinkel, Y. (2020). The Levant in the Pottery Neolithic and Chalcolithic Periods. In K. Radner, N. Moeller, & D. T. Potts (Eds.), *The Oxford History of the Ancient Near East – Volume I: From the Beginnings to Old Kingdom Egypt and the Dynasty of Akkad* (pp. 1439–61). Oxford: Oxford University Press.

Garrard, A., Baird, D., Colledge, S., Martin, L., & Wright, K. (1994). Prehistoric Environment and Settlement in the Azraq Basin: An Interim Report on the 1987 and 1988 Excavation Seasons. *Levant*, 26, 73–109.

Goring-Morris, A. N., & Belfer-Cohen, A. (2020). Highlighting the PPNB in the Southern Levant. *Neo-lithics*, 20, 3–22.

Hole, F., & Flannery, K. (1968). The Prehistory of Southwestern Iran: A Preliminary Report. *Proceedings of the Prehistoric Society*, 33, 147–206.

Hole, F., Flannery, K. V., & Neely, J. A. (1969). *Prehistory and Human Ecology of the Deh Luran Plain; an Early Village Sequence from Khuzistan, Iran. Memoirs of the Museum of Anthropology, University of Michigan, no. 1.* Ann Arbor: Museum of Anthropology, University of Michigan.

Horejs, B., Milić, B., Ostmann, F., Thanheiser, U., Weninger, B., & Galik, A. (2015). The Aegean in the Early 7th Millennium BC: Maritime Networks and Colonization. *Journal of World Prehistory*, 28(4), 289–330. doi:10.1007/s10963-015-9090-8

Huot, J.-L. (1992). The First Farmers at Oueili. *Biblical Archaeologist*, 55(4), 188–95.

Huot, J.-L. (1996). *Oueili, Travaux de 1987 et 1989.* Paris: Centre National de la Recherche Scientifique.

Khalaily, H., Re'em, A., Vardi, J., & Milevski, I. (2020). *The Mega Project at Motza (Moża): The Neolithic and Later Occupations up to the 20th Century.* Jerusalem: Israel Antiquities Authority.

Kirkbride, D. (1974). Umm Dabaghiyah: A Trading Outpost? *Iraq*, 36(1–2), 85–92. doi:10.2307/4199976

Köhler-Rollefson, I. (1992). A Model for the Development of Nomadic Pastoralism on the Transjordan Plateau. In O. Bar-Yosef & A. Khazanov (Eds.), *Pastoralism in the Levant: Archaeological Materials in Anthropological Perspective* (Vol. Monographs in World Prehistory, No. 10, pp. 11–8). Madison, WI: Prehistory Press.

Lloyd, S., Safar, F., & Braidwood, R. (1945). Tell Hassuna Excavations by the Iraq Government Directorate General of Antiquities in 1943 and 1944. *Journal of Near Eastern Studies*, 4(4), 255–89.

Marciniak, A. (2015). A New Perspective on the Central Anatolian Late Neolithic. The TPC Area Excavations at Çatalhöyük. In S. R. Steadman & G. McMahon (Eds.), *The Archaeology of Anatolia. Recent Discoveries (2011–2014)* (Vol. I) (pp. 6–25). Newcastle upon Tyne: Cambridge Scholars Publishing.

Merpert, N., & Munchaev, R. (1987). The Earliest Levels at Yarim Tepe I and Yarim Tepe II in Northern Iraq. *Iraq*, 49, 1–36. doi:10.2307/4200262

Miller, H., Baird, D., Pearson, J., Lamb, A. L., Grove, M., Martin, L., & Garrard, A. (2019). The Origins of Nomadic Pastoralism in the Eastern Jordanian Steppe: A Combined Stable Isotope and Chipped Stone Assessment. *Levant*, 50(3), 1–24. doi:10.1080/00758914.2019.1651560

Molleson, T. (1994). The Eloquent Bones of Abu Hureyra. *Scientific American*, 271(2), 70–5.

Molleson, T. (2000). The People of Abu Hureyra. In A. M. T. Moore, G. C. Hillman, & A. J. Legge (Eds.), *Village on the Euphrates: From Foraging to Farming at Abu Hureyra* (pp. 301–24). Oxford: Oxford University Press.

Moore, A. M. T., Hillman, G. C., & Legge, A. J. (2000). *Village on the Euphrates: From Foraging to Farming at Abu Hureyra*. Oxford: Oxford University Press.

Nieuwenhuyse, O. P., Roffet-Salque, M., Evershed, R. P., Akkermans, P., & Russell, A. (2015). Tracing Pottery Use and the Emergence of Secondary Product Exploitation through Lipid Residue Analysis at Late Neolithic Tell Sabi Abyad (Syria). *Journal of Archaeological Science*, 64, 54–66. doi:10.1016/j.jas.2015.10.002

Oates, J. (1969). Choga Mami, 1967–68: A Preliminary Report. *Iraq*, 31(2), 115–52. doi:10.2307/4199877

Özdoğan, A. (2011). Çayönü. In M. Özdoğan, N. Başgelen, & P. Kuniholm (Eds.), *The Neolithic in Turkey. New Excavations and New Research - The Tigris Basin* (pp. 185–269). Istanbul: Arkeoloji ve Sanat Yayinlari.

Rollefson, G. (2020). The PPNC: LIke a Bridge over Troubled Waters. In H. Khalaily, A. Re'em, J. Vardi, & I. Milevski (Eds.), *The Mega Project at Motza (Moża): The Neolithic and Later Occupations up to the 20th Century* (pp. 131–62). Jerusalem: Israel Antiquities Authority.

Rollefson, G., Athanassas, C., Rowan, Y., & Wasse, A. (2016). First Chronometric Dates for "Works of Old Men": Late Prehistoric "Wheels" Near Wisad Pools, Black Desert, Jordan. *Antiquity*, 90(352), 939–52.

Rollefson, G., Rowan, Y., & Wasse, A. (2014). The Late Neolithic Colonization of the Eastern Badia of Jordan. *Levant*, 46(2), 285–301.

Rollefson, G. O. (2011). The Greening of the Badlands: Pastoral Nomads and the "Conclusion" of Neolithization in the Southern Levant. *Paléorient*, 37(1), 101–9.

Rowan, Y., Rollefson, G., Wasse, A., Abu-Azizeh, W., Hill, A. C., & Kersel, M. M. (2015). The "Land of conjecture:" New Late Prehistoric Discoveries at Maitland's Mesa and Wisad Pools, Jordan. *Journal of Field Archaeology* 40(2). doi:10.1179/0093469015Z.000000000117

Sherratt, A. (1981). Plough and Pastoralism: Aspects of the Secondary Products Revolution. In I. Hodder, G. L. Isaac, & N. Hammond (Eds.), *Pattern of the Past* (pp. 261–306). Cambridge: Cambridge University Press.

Weninger, B., Clare, L., Gerritsen, F. A., Horejs, B., Krauß, R., Linstädter, J., . . . Rohling, E. J. (2014). Neolithisation of the Aegean and Southeast Europe during the 6600–6000 cal BC Period of Rapid Climate Change. *Documenta Praehistorica*, 41, 1–31.

Weninger, B., Clare, L., Rohling, E. J., Bar-Yosef, O., Böhner, U., Budja, M., . . . Zielhofer, C. (2009). The Impact of Rapid Climate Change on Prehistoric Societies during the Holocene in the Eastern Mediterranean. *Documenta Praehistorica*, 36, 7–59.

Wilkinson, T. J., & Tucker, D. J. (1995). *Settlement Development in the North Jazira, Iraq: A Study of the Archaeological Landscape*. Warminster: Aris & Phillips for the British School of Archaeology in Iraq and the Dept. of Antiquities & Heritage, Baghdad.

Yoffee, N., & Clark, J. J. (1993). *Early Stages in the Evolution of Mesopotamian Civilization: Soviet Excavations in Northern Iraq*. Tucson: University of Arizona Press.

9 The Evolutionary Framework for the Story

Our ENT is an episode in the long term of human evolutionary history; for many readers interested in the archaeology of the ENT the recent work on cultural evolution, cultural niche construction theory and cultural cumulation may be unfamiliar, but these ideas are the underpinning of the account that I will unfold in the next chapter. This unfamiliar underpinning needs introducing in a chapter of its own.

I am not attempting to show why I think that a cultural evolutionary account is the best way of explaining how and why the transformation occurred, and I am not engaging in an historical discussion of the various theoretical frameworks that have been proposed to account for the Neolithic, starting with Gordon Childe's Neolithic revolution, and progressing through all the arguments for and against successive theoretical debates. That might be the necessary first section if this were a PhD thesis. Such a critical discussion would be a lengthy digression from the objective of this book. Rather, I propose to explain briefly why I personally have been uncomfortable with what has been the orthodox kind of account over the many years that I have been teaching and researching, before turning to the subject of cultural evolution, which seems to me the way to develop an understanding of our period.

There is simple reason why we have not been able to generate a satisfactory explanation of the processes that produced the Neolithic: we have, from the beginning, been asking the wrong or tangential questions. In the first place, dating back to Braidwood's pioneering inter-disciplinary research programme of the 1950s and 1960s, research interest became closely focused on the identification of the geographical region where, and the date when, plants and animals were first domesticated. Secondary to that has been the quest for a climatic and environmental event that could be the factor driving the human adaptations that make the Neolithic. Childe's hypothesis was exactly that, a speculative hypothesis (Childe 1936); after the second world war, Robert Braidwood set out to test his significantly modified version of Childe's hypothesis in the field laboratory of 'the hilly flanks of the Fertile Crescent'. Childe had written with rather overblown rhetoric in his popular paperback *What Happened in History* that the Neolithic revolution marked an escape from 'the impasse of savagery' making its 'participants active partners with nature instead of parasites on nature' (Childe 1942: 55). Braidwood recognised that, whatever the revolutionary impact of Neolithic agriculture and herd management, there was a pre-Neolithic process whereby the classic Upper Palaeolithic peoples living

DOI: 10.4324/9781351069281-10

in the hilly regions where the progenitors of Neolithic crops and herds existed in the wild. However, his inter-disciplinary field research could not find evidence of a potential climatic event that might have driven the transition from hunting and gathering to cultivation and herding. Much of his good work was side-lined when the younger generation of 'processualist' anthropologist-archaeologists took centre stage in the mid-1960s. Lewis Binford (1968) followed by Kent Flannery (1969) proposed different versions of an environmental-ecological process which cited the pressure of expanding hunter-gatherer populations on available wild food resources as the driver of adaptations. Flannery's modelling of his proposed 'broad-spectrum revolution' was based on his knowledge of the Epipalaeolithic and Neolithic, and his fieldwork experience with Frank Hole, another of Braidwood's former PhD students, in southwest Iran. If there was no sign of a climatic-environmental down-turn that reduced the available resources for hunter-gatherers, then the equation could be turned the other way around: levels of human population grew beyond the capacity of available natural food resources.

The question was reversed again, as the Younger Dryas phase came into fo-cus. Gordon Hillman and his research students, analysing in meticulous detail the plant remains from the final Epipalaeolithic occupation at Abu Hureyra, proposed that the small changes in the proportions of different species could be interpreted as the beginnings of the cultivation of rye, necessitated by the reduction in avail-able resources caused by the cruel onset of the cooler, drier climatic conditions (Hillman et al. 2001; Moore & Hillman 1992). There has been much further dis-cussion of the impact of the Younger Dryas in southwest Asia, which has gener-ally diminished its effects; indeed, if has been said that the Younger Dryas may have actually increased the availability of some wild food resources in s some parts of the region.

The problem is that neither version of this kind of human adaptive-response-to-environmental-pressure account works: the theory is deficient, and in any case the prehistoric facts point to a different story. This was at the heart of Jacques Cau-vin's rejection of the common idea that environmental pressures, whether through declining natural food resources or through increasing population, had led to the Neolithic and the beginning of farming (Cauvin 1994, 2000). Cauvin's proposed solution was exciting and challenging (and I cannot do it justice in two or three simple sentences): people in southwest Asia had taken up sedentism, and the new way of life provoked what he called a 'psycho-cultural' revolution, which we see in the earliest Neolithic explosion of striking visual imagery. Armed with this new perception of the world and their place within it, people began to take control of plants and animals. A little before Cauvin's book, and shortly before he began his 25-year project at Çatalhöyük, Ian Hodder (1990) published his intriguingly differ-ent take on domestication and the Neolithic beginnings of the *domus* (= house or home, the Latin root of the word), suggesting that the first settlements with build-ings that we can recognise as houses were a new way of living together, with its own norms and rituals. Around this time, too, it was beginning to become clear that domestication, whether of plants or animals, was not an event, or a short process, accomplished in a few generations.

Having been a vaguely uncomfortable bystander as the contest between the Younger Dryas or population pressure went on, the salvage excavation of Qermez Dere in northern Iraq in the 1980s was a turning point for me. We learned that the settlement was in place at the boundary between the end of the Younger Dryas and the very beginning of the Holocene, at a time and at a place – far from the supposed Levantine refuge - where Hillman's reconstruction of the impact of the Younger Dryas made it next to impossible for sedentary hunter-gatherers. The experience of excavating the extraordinary subterranean buildings and their (irrational) destructions and replacements (Watkins 1990), closely followed by reading those books by Hodder and Cauvin, spurred new ideas. I had the good fortune to be able to work closely with Jacques Cauvin and his thought for more than a year as I produced the English language version of his major book (Cauvin 2000). I have constantly regretted that Jacques' death at the time that the English edition appeared prevented me from pursuing the dialogue in person that I had enjoyed while battling with his prose. Those authors introduced quite different perspectives on the neolithisation process. While both these books pointed to the vivid and powerful imagery, symbolism and rituals that preceded or accompanied the move to agriculture, giving us new ways to fit the Neolithic world together, they did not explain how this Neolithic world and its fascination with symbolic representation came to be.

At much the same time, Steven Mithen's (1996) *The Prehistory of the Mind* pioneered the linking of material culture to new theories on the evolution of human cognition. In the same decade Robin Dunbar burst onto the scene with his popular book *Grooming, Gossip and the Evolution of Language* (Dunbar 1997), and his 'social brain hypothesis' (Dunbar 1998). It was through Dunbar's collaboration with Clive Gamble and John Gowlett in the Lucy to Language project that his ideas came to be worked out against the archaeological evidence (Dunbar et al. 2010a, 2010b, 2014; Gamble et al. 2014). Through cognitive psychology and the cognitive capacities of the human brain, Dunbar's social brain hypothesis links biological evolution in the shape of the physical and cognitive evolution of the human brain with the evolution of human society and the size and cohesiveness of social groups. These were researchers who were trialling new ways of applying evolutionary theory in archaeology, and new advances in evolutionary theory.

Advances in Cultural Evolutionary Theory: From Gene-Culture Evolution to Cultural Niche Construction

It is important to recognise that, since the middle of the twentieth century, evolutionary theory has developed and expanded to become much richer under the heading of the 'extended evolutionary synthesis'. I was impressed by Robert Braidwood's question: if the beginnings of cultivation and animal herding were indeed an adaptive response to climate change and environmental deterioration, why had it not happened at one of the many earlier Pleistocene climatic oscillations; why not at an earlier Pleistocene climate oscillation? Braidwood had finally concluded that perhaps culture was just not ready (Braidwood & Willey 1962: 342). Over the last twenty or so years, there have emerged and solidified advances in evolutionary theory that make it possible for us to frame a new account of the ENT process in terms that are now established elements of the extended evolutionary synthesis, in particular, cultural niche construction theory and cumulative culture. Perhaps it was only at the end of the Palaeolithic and the beginning of the Neolithic that cultural conditions were ready.

In the middle of the twentieth century, the new science of genetics and the discovery of DNA seemed to provide solutions to the missing element of Darwinian evolution, the mechanism of inheritance. This was 'the modern synthesis' (Huxley 1942) that was expounded and developed in books like *The Selfish Gene* (1976) and *The Extended Phenotype* (1982) that made Richard Dawkins's name famous. Those advances opened the way for further developments in evolutionary theory. In 2015 eight of the leading scientists and thinkers involved in different branches of evolutionary theory co-authored a key paper published in the *Proceedings of the Royal Society* (Laland et al. 2015). They spelled out the structure, core assumptions and novel predictions of 'the extended evolutionary synthesis' or EES. In a fascinating book-length discussion of important elements of the EES, Eva Jablonka and Marion Lamb write about the new way of understanding evolutionary theory in terms of four dimensions of evolution: genetic, epigenetic, behavioural, and symbol-based (Jablonka & Lamb 2005, 2007). The developments in the behavioural and symbol-based fields amount to cultural evolution, which is what concerns us.

There are three elements that we need for building an account of cultural evolution: gene-culture co-evolution, cultural niche construction theory, and cumulative culture. Gene-culture co-evolution began to appear in the literature more than 20 years ago (Feldman & Laland 1996). The term refers to the interactive evolutionary process whereby cultural processes shape genetic evolution by modifying the selection of genes, which often entails reciprocal interactions and feedbacks reinforcing or refining the cultural processes. These and the further developments of cultural evolutionary theory, it should be said, have proved controversial both for some biologists on the one hand, and on the other hand for historians and some prehistorians. We have already encountered gene-culture co-evolution in Dunbar's social brain hypothesis, although Dunbar himself does not use the term. In his social brain hypothesis, in environments where group size is an ecological advantage, coping with the increasingly complex social relations among the group's members favours those with more brainpower and a larger brain, which in turn favours further increases in social group size. Boyd et al. (2010) have given us a useful survey of the subject with some relevant examples; Liu and Stout (2022) open their paper on the cultural reproduction of lithic technology with a useful introduction to the subjects of cultural evolution and cumulative culture. The case has recently been argued for the co-evolution of human language and the brain's capacity to learn language, leading to the progressive upgrading of the complexity of language matched by brain evolution (Dor & Jablonka 2014; Kirby et al. 2007). Language is of course of great importance for human communication, whether it is spoken or written; but, as written language records and stores information, so do other modes of symbolic communication that humans have evolved, which will be of importance in the next chapter.

These co-evolutionary feedback loops (labelled 'reciprocal causation' by evolutionary scientists) involve a closed-circuit interaction with one another in which each responds to the other. The most obviously relevant example of gene-culture co-evolution for us is the domestication of plants and animals (Smith 2016; Zeder 2017). The cultivation of selected species in tended plots led to morphological and genetic changes so that wheat and barley, for example, became essential food resources for humans, while those cereals came to rely on humans sowing their seeds and were no longer self-reproducing.

While each of those developments can be seen as examples of reciprocal causation, the picture becomes less simple as they enmesh with each other, prompting further co-evolution feedback loops to emerge. Cultural systems are obviously complex systems, whose various components interact with one another and are dependent on one another. Without using the terms that are peppering these paragraphs, the complexity of interacting elements is what has excited Ian Hodder (2011, 2016) in his discussions of 'entanglement'. An important feature of complex systems theory is that in a complex system, a cultural system, or a cultural niche, no single component can be identified as causal: the system evolves dynamically as its many components interact with one another. We can see how some of the various components interact in feedback loops, and we can see how developments in one area of the system can have knock-on effects on other areas; but we cannot usually identify one component as the single causative factor.

And that leads us to most important component of the extended evolutionary synthesis, niche construction theory. The evolutionary biologist Kevin Laland, together with Professor John Odling-Smee, first proposed the theory (Odling-Smee et al. 1996, 2003). The simplest definition of the term appears in a paper in which Laland and archaeologist Michael O'Brien set out to explain the significance of (cultural) niche construction for archaeology: 'niche construction is 'the capacity of organisms to modify natural selection in their environment and thereby act as co-directors of their own, and other species', evolution' (Laland & O'Brien 2010: 303). Niche construction exists throughout the biological world, among many animals which make nests, burrows, webs, and pupal cases, and including plants that change levels of atmospheric gases and modify nutrient cycles, as well as fungi that decompose organic matter, and bacteria that fix nutrients. Humans have become the most active niche constructors because of their capacity for culture. Humans operate within niches which they themselves have formed, and which become the effective environment that accommodates them and to which they accommodate (Laland et al. 2001; Laland & O'Brien 2011). Botanists and zoologists are interested in the backwards and forwards interaction between human practices and the species that humans have taken into their cultural niche with domestication. Psychologists are equally interested in the ways that the humanly constructed cultural niche in turn affects the cognitive functioning of its builders. Linguists interested in the evolution of language, for example, juggle with the co-evolution of the unique human vocal tract, the cognitive capacity for a theory of mind (which allows us to take into consideration the situation of the person we are speaking to as we plan what we want to say to them), and the cognitive capacity to attribute significance and meaning to symbols such as words. The archaeologist Dietrich Stout (2011, 2019) discusses the co-evolution of Lower Palaeolithic stone tool-making and brain size, bringing into interaction the technical skills of stone toolmaking, emergent language skills, and the cognitive skills involved in teaching and learning. Antón et al. (2014) think of the complex web of interactions within the cultural niche of the earliest Homo, bringing together increasing brain size, increased tool-making skills, transport from a distance of quality raw material for stone tools, expansion of diet, and greater developmental plasticity (the capacity to adjust to different or changing environmental conditions), which expands the potential range of human populations. The human cultural niche has been a complex web of interacting elements from an early stage in human evolution: our contemporary cultural niche is of bewildering complexity, and few of us would survive for more than a few days if we were to be shipwrecked like Robinson Crusoe on a desert island.

 Very quickly the biologists and evolutionary scientists who were key in the formulation of niche construction theory realised that it applied very significantly to humans; hence they have enthusiastically explored the potential of cultural niche construction theory (Laland 2017; Laland et al. 2001, 2016; Laland & O'Brien 2010, 2011). Kim Sterelny employs cultural niche construction theory in his major book (*The Evolved Apprentice: How Evolution Made Humans Unique*. 2011), describing how humans have engaged in 'ecological engineering' to create an 'ecological inheritance', but have also engaged in 'epistemic engineering' to create 'epistemic inheritance'. 'Ecological inheritance' refers to the constructed selective environment within which the next and succeeding generations will live, supporting long childhoods, for example, which provide the time and opportunities for the extended social learning required for human social life. 'Epistemic inheritance' refers to the special form of niche construction that facilitates cultural learning and the transmission of cultural knowledge and skills. For at least half of the existence of Homo sapiens, as Kendal, Tehrani and Odling-Smee (2011: 980) argue:

> humans have become increasingly reliant on physical and semantic resources [that is, bodies of cultural knowledge that are transmitted and learned] that have been shaped by the cultural activities of preceding generations — from domesticated animals and tool-making to writing, the built environment and even religious cosmologies. Such inventions have in various ways both depended on and then subsequently shaped the evolution of genetic and other cultural traits.

Over the long term of human evolution humans have constantly developed and further developed their unique human cultural niche. Relatively early in human evolution, for example, the control of fire enabled forager groups to protect their home-base against the cold, and make it more secure from predators at night (Sánchez-Villagra & van Schaik 2019; Theofanopoulou et al. 2017). It also offered the facility to cook food, resulting in a more efficient digestion of nutritional resources, which released more energy resources to feed energy-demanding brainpower (Aiello & Wheeler 1995). Cultural niche construction theory was proposed at the time that they were publishing their paper on what they called 'the expensive tissue hypothesis', but it neatly illustrates the interaction of multiple different feedback loops within the cultural niche. Aiello and Dunbar could therefore argue for the increasing neo-cortex ratio among the hominins, fuelled by improvements in diet and a less demanding digestive system, supporting the emergence of language (which allows groups of individuals to chat and gossip) replacing one-to-one grooming in increasingly large social groups (Aiello & Dunbar 1993). The cultural niche is much more than a gene-culture co-evolutionary feedback loop; it is a complex system of many and diverse interconnected biological and cultural feedback loops.

I said that there were three new evolutionary features; I have said a little about gene-culture co-evolution, and more about cultural niche construction. The third mechanism, cumulative cultural evolution, seems to be distinctive of humans and their culture (although there are claims for limited forms of cumulative culture in a variety of animal species, see, e.g., Reader & Laland 2002). Over the two or more millions of years of evolutionary time, humans have gradually evolved uniquely efficient modes of social learning. Social learning means much more than simply learning by imitation. The would-be learner must be able to take a perspective on what he or she wishes to learn, involving understanding what is the objective in the mind of the person from whom she is learning. The same applies in the opposite direction for the 'expert' imparting the skill and its attendant knowledge to the learner; it pays for the teacher to have in mind the inexpert situation of the person who is seeking to learn. Of both partners in the teaching-learning process it requires what psychologists call theory of mind. Over evolutionary time human societies have become so organised that they have the capacity to accurately transmit extensive and diverse bodies of their cultural knowledge and skills across the generations. Furthermore, human societies have evolved—very slowly at first but with increasing efficiency—a complex of skills and modes of social life that enable them not only to retain their store of cultural knowledge but also to accumulate cultural innovations (Mesoudi & Thornton 2018). This evolved capacity to accumulate cultural innovations is particularly important for us in the ENT (not to mention the importance of formal education, technical innovation and the deliberate expansion of knowledge in our present world).

Before thinking about innovation and the expansion and improvement of elements of their culture, it is of crucial importance that a social group can maintain and transmit to the next generation their current and inherited cultural package of skills, practices, conventions, and knowledge. Kim Sterelny (2011) traces the long-term development of human cooperation and the evolution of social and cognitive skills embedded in a cultural niche adapted for cultural transmission. Certainly by the time of Homo sapiens, juveniles and adolescents living in extended communities relied less on parental teaching and were adept at identifying their role models and the best teachers from whom to learn advanced cultural skills; and there were cultural norms that enabled skilled and experienced older people to work with young people to transmit their skills—what Sterelny calls apprentice learning.

Recent anthropological and experimental work shows that small-scale foraging band societies are subtly structured to maximise inter-connections between bands and interactions between non-related individuals or groups (Derex & Mesoudi 2020, which gives a detailed and up-to-date survey with plentiful references). Analysis shows that much innovation involves the refinement of existing things or the recombination of existing elements, rather than outright invention. In order to improve the chances of the emergence of innovations it is therefore important in band-level societies to maximise the ways that individuals from one group encounter people in other groups. Recent experimental work has shown that the best environment for transferring knowledge or encouraging innovation is to set a task to several small groups of people, but to allow individuals to move between groups, making observations and comparing notes, as it were, and thus generating insights. These experimental groups closely mirror the social structures identified in contemporary hunter-gatherer societies, suggesting that hunter-gatherer societies have evolved forms of cultural niche that best fit them for their lives as small-scale mobile bands whose members sometimes visit other bands, or move from one band to another (Hill et al. 2014).

On the basis of decades of collaborative research in evolutionary theory, ethnographic fieldwork and laboratory experiments, the American economist and anthropologist Joe Henrich argues, like Sterelny, for the power of the cultural learning niche for the safe inter-generational transfer of complex knowledge and diverse skills (Henrich 2015). Henrich promises in the subtitle of his book to show 'How Culture Is Driving Human Evolution, Domesticating Our Species, and Making Us Smarter'. In short, humans have evolved to become more and more social, not just in terms of the scale of human societies, but more importantly in terms of the complexity and intensity of our social relations with one another and our capacities for cooperation; we have been described as pro-social (e.g. Boyd & Richerson 2009), or even ultra-social (Tomasello 2014; Turchin 2013).

The advance that has above all characterised recent human evolution is the human capacity for 'cumulative cultural evolution'. The accumulation of innovation within a cultural niche is in turn dependent on the existence of very cohesive social groups and a cultural niche that provides for the tutoring, acquisition and practice of complex skills: that is what both Sterelny and Henrich discuss extensively. There is a demographic component in this advanced cultural learning niche: when a group relies on complex and compound cultural skills and know-how, for example to build a kayak, make a harpoon, and hunt seals in the Arctic Ocean, there must be a population large enough for there to be several wise and experienced practitioners of the complex skills. In such conditions the complex bodies of knowledge, skill and lore can be retained and transmitted; and, from time to time, someone will introduce a small improvement, which others will notice and adopt – cultural accumulation.

Upper Palaeolithic societies that were comprised of small, scattered, forager bands were very successful, but they were at the limits of their cultural capacity to sustain sufficiently large numbers of people who could maintain meaningful contact with one another. Henrich provides examples of small-scale societies which, for whatever reason, were reduced in number or became isolated, so that they progressively lost some elements of their cultural package. He encapsulates his overall conclusion thus: from an early stage in the human evolutionary story, 'cultural evolution became the primary driver of our species' genetic evolution' (Henrich 2015: 57). Henrich describes the accelerating process of interactions as 'autocatalytic', meaning that it produces the fuel that propels itself – in short, it becomes a runaway process (cf. Rendell et al. 2011). In a recently published advance of that line of research, a group of researchers, including Henrich, modelled and simulated the long-term process of human biological, cognitive, social and cultural co-evolution under the label 'the cumulative cultural brain hypothesis' (Muthukrishna et al. 2018). Their simulation explores the complex of components that precipitated and sustained 'an autocatalytic take-off' in the long-term trajectory of human cultural, cognitive, social and biological evolution. Laland summarises his central argument thus:

> no single prime mover is responsible for the evolution of the human mind. Instead, I highlight the significance of accelerating cycles of evolutionary feedback, whereby an interwoven complex of cultural processes to reinforce each other in an irresistible *runaway* [my italics] dynamic that engineered the mind's breathtaking computational power.
>
> (Laland 2017: 3)

The word 'autocatalytic' appears repeatedly, too, in that research report: the interactions between the expansion in scale and complexity of cultural knowledge, of social group size and of the intensity and efficiency of social interconnectedness over time became a self-driving, runaway process.

Henrich is an anthropologist with a deep knowledge of economics; Sterelny is a philosopher. Laland has developed his research from a starting point in biology. In his recent book, *Darwin's Unfinished Symphony* (2017), Laland summarises over 20 years of collaborative research; in the introductory chapter he writes of the 'unprecedented interaction between cultural and genetic processes in human evolution' that have fuelled 'a relentless acceleration', an 'autocatalytic process ... with accelerating cultural change driving technological progress and diversification' (Laland 2017: 29). There once again is that word 'autocatalytic', defining the niche conditions of accelerating cumulative cultural evolution as self-fuelling.

A demographic factor is important to understanding our ENT, as we shall see in the following chapter. Henrich, Laland and Sterelny show how important the maintenance of large populations has been for sustaining increasingly complex human cultures. Comparing the consistent wealth of symbolic representations of different kinds found in the Upper Palaeolithic of Europe beginning around 45,000 years ago with the transient record of symbolic activity among their Middle Stone Age predecessors in southern Africa, another group of researchers has concluded that the sustaining and growth of cultural complexity from the Upper Palaeolithic onwards depended on demographic factors (Powell et al. 2009). In the following chapter I will argue that the establishment of permanently co-resident communities within the ENT was the emergence of a new kind of cultural niche that enabled the concentration of population numbers and dense webs of social interaction; in short, the ENT was the formation of the cultural niche that imparted a sharp up-kick to that runaway, autocatalytic process and enhanced the capacity for cultural accumulation.

And Extended Minds

There is a second area of recent debate that needs to be introduced, because it also comes into play in the following chapter. It is convenient to take a 1998 paper as the starting point, because it is referred to by almost everyone who is engaged with this subject, and its title, 'The Extended Mind' (Clark & Chalmers 1998), has become the general label for the field. The idea of the extended mind has gathered much interest, shown for example in the book of collected essays edited by Richard Menary (2010), and numerous papers in journals that may be unfamiliar to most archaeology readers. That original paper seems dated now, as it was written when the worldwide web was very young, and before mobile phones became pervasive and practically indispensable. It introduced the importance of material scaffolding to our human cognition. Clark (1997, 1998, 2007, 2008) has explored how we have taken to using 'stuff to make us smarter' (to borrow a phrase from Kim Sterelny), so that some of these physical aids to our cognition should be understood as parts of our minds, and certainly parts of our memory. We offload a great deal of information that is important to us into notebooks or the contacts and calendar sections of our mobile phones; they become forms of external memory, and we can cope with much more if we don't have to carry everything in our heads.

Kim Sterelny uses that Clark & Chalmers paper as the starting point of his essay on 'the archaeology of the extended mind' (Sterelny 2019). While Clark introduced the idea of the extended mind with illustrations of the way that an address book acts as an extension of the owner's memory, Sterelny brings out other ways in which material things serve as shared memory aids when experienced elders are teaching complex knowledge and a difficult skill to juniors. More importantly, he discusses the making of cultural things that act as signals or material symbols whose meanings or significance can be 'read' by anyone within their cultural orbit. Clark was concerned with the individual's physical memory extensions. Like Sterelny, I would emphasise the way that we all 'read' shared material signals and symbols, whether it is common road-signs or the architecture or a great cathedral or mosque.

Here we are entering a field where the neuropsychologist Merlin Donald has established the groundwork with his theory of the three-stage evolution of human communication and cognition (Donald 1991, 2001). The first stage in Donald's scheme is 'mimetic', where communication is through sounds and gestures; the second stage, which involves language, he labels 'mythic' because of the capacity for narrative and story-telling. The ultimate form of Donald's third stage is writing, which expands on the power of spoken language; Donald identifies this power as 'extended symbolic storage', which had begun to emerge in the Palaeolithic from before 100,000 years ago in the form of things like beads and personal ornaments, and from around 45,000 years ago in art and sculpture, but which was expanded within our period of the ENT (and in its equivalent periods in other parts of the world) in the form of architecture and various forms of elaborate material symbolism and symbolic action. Colin Renfrew recognised the significance of Merlin Donald's work and organised a conference in Cambridge at which Donald was the keynote speaker. Donald's main paper, complemented by a second paper with his reflections in consequence of his encounter with the contributing archaeologists, discuss the relationship between his theory and the archaeology of human evolution (Donald 1998a, 1998b). There is more to be said on this subject in the following chapter, as Donald's idea of the evolution of modes of 'extended symbolic storage' is singularly relevant to the monumental architecture, symbolic sculpture, and small-scale figurines of our Neolithic (Watkins 1990, 2004a, 2004b, 2012, 2017). Here it is enough to say that developments of the extended or distributed mind, together with the recent developments in cultural evolutionary theory, play key roles in working out the significance of our Epipalaeolithic–Neolithic transformation in the next chapter.

References

Aiello, L., & Dunbar, R. I. M. (1993). Neocortex Size, Group Size and the Evolution of Language. *Current Anthropology*, 36, 184–93.

Aiello, L. C., & Wheeler, P. (1995). The Expensive Tissue Hypothesis: The Brain and the Digestive System in Human and Primate Evolution. *Current Anthropology*, 36(2), 199–221.

Antón, S. C., Potts, R., & Aiello, L. C. (2014). Evolution of Early Homo: An Integrated Biological Perspective. *Science*, 345(6192). doi:10.1126/science.1236828

Binford, L. R. (1968). Post-Pleistocene Adaptations. In L. R. Binford & S. R. Binford (Eds.), *New Perspectives in Archaeology* (pp. 313–41). Chicago, IL: University of Chicago Press.

Boyd, R., & Richerson, P. J. (2009). Culture and the Evolution of Human Cooperation. *Philosophical Transactions of the Royal Society B-Biological Sciences*, 364(1533), 3281–8. doi:10.1098/rstb.2009.0134

Boyd, R., Richerson, P. J., & Henrich, J. (2010). Gene-Culture Coevolution in the Age of Genomics. *Proceedings of the National Academy of Sciences of the United States of America*, 107(Supplement 2), 8985–92.

Braidwood, R. J., & Willey, G. R. (1962). Conclusions and Afterthoughts. In G. R. Willey & R. J. Braidwood (Eds.), *Courses towards Urban Life: Archaeological Considerations of Some Cultural Alternatives* (pp. 330–59). Chicago, IL: Aldine Publishing Company.

Cauvin, J. (1994). *Naissance des divinités, naissance de l'agriculture: la révolution des symboles au Néolithique*. Paris: CNRS editions.

Cauvin, J. (2000). *The Birth of the Gods and the Origins of Agriculture*. Cambridge: Cambridge University Press.

Childe, V. G. (1936). *Man Makes Himself*. London: C. A. Watts.

Childe, V. G. (1942). *What Happened in History*. Harmondsworth: Penguin Books.

Clark, A. (1997). *Being There: Putting Brain, Body, and World Together Again*. Cambridge; London: MIT Press.

Clark, A. (1998). Where Brain, Body and World Collide. *Daedalus*, 127(2), 257–80.

Clark, A. (2007). Re-Inventing Ourselves: The Plasticity of Embodiment, Sensing, and Mind. *Journal of Medicine and Philosophy*, 32(3), 263–82. doi:10.1080/03605310701397024

Clark, A. (2008). *Supersizing the Mind: Embodiment, Action, and Cognitive Extension*. Oxford; New York, NY: Oxford University Press.

Clark, A., & Chalmers, D. J. (1998). The Extended Mind. *Analysis*, 58(1), 7–19.

Dawkins, R. (1976). *The Selfish Gene*. Oxford: Oxford University Press.

Dawkins, R. (1982). *The Extended Phenotype*. Oxford: Oxford University Press.

Derex, M., & Mesoudi, A. (2020). Cumulative Cultural Evolution within Evolving Population Structures. *Trends in Cognitive Sciences*, 24(8), 654–67. doi:10.1016/j.tics.2020.04.005

Donald, M. (1991). *Origins of the Modern Mind: Three Stages in the Evolution of Culture and Cognition*. Cambridge, MA; London: Harvard University Press.

Donald, M. (1998a). Hominid Enculturation and Cognitive Evolution. In C. Renfrew & C. Scarre (Eds.), *Cognition and Material Culture: The Archaeology of Symbolic Storage* (Vol. McDonald Institute monographs, pp. 7–17). Cambridge: McDonald Institute for Archaeological Research.

Donald, M. (1998b). Material Culture and Cognition: Concluding Thoughts. In C. Renfrew & C. Scarre (Eds.), *Cognition and Material Culture: The Archaeology of Symbolic Storage* (pp. 181–7). Cambridge: McDonald Institute for Archaeological Research.

Donald, M. (2001). *A Mind So Rare: The Evolution of Human Consciousness*. New York, NY: Norton.

Dor, D., & Jablonka, E. (2014). Why We Need to Move from Gene–Culture Co-Evolution to Culturally-Driven Co-Evolution. In D. Dor, K. Knight, & J. Lewis (Eds.), *The Social Origins of Language* (pp. 15–30). Oxford: Oxford University Press.

Dunbar, R. I. M. (1997). *Grooming, Gossip and the Evolution of Language*. London: Faber.

Dunbar, R. I. M. (1998). The Social Brain Hypothesis. *Evolutionary Anthropology*, 6(3), 178–90. doi:10.1002/(SICI)1520-6505(1998)6:5<178

Dunbar, R. I. M., Gamble, C., & Gowlett, J. A. J. (2010a). The Social Brain and the Distributed Mind. *Proceedings of the British Academy*, 158, 3–15.

Dunbar, R. I. M., Gamble, C., & Gowlett, J. A. J. (2010b). *Social Brain, Distributed Mind*. Oxford: Oxford University Press & British Academy.

Dunbar, R. I. M., Gamble, C., & Gowlett, J. A. J. (2014). *Lucy to Language: The Benchmark Papers*. Oxford: Oxford University Press.

Feldman, M. W., & Laland, K. N. (1996). Gene-Culture Coevolutionary Theory. *Trends in Ecology & Evolution*, 11(11), 453–7.

Flannery, K. V. (1969). The Origins and Ecological Effects of Early Domestication in Iran and the Near East. In P. J. Ucko & G. W. Dimbleby (Eds.), *The Domestication and Exploitation of Plants and Animals* (pp. 73–100). London: Duckworth.

Gamble, C., Gowlett, J., & Dunbar, R. (2014). *Thinking Big: How the Evolution of Social Life Shaped the Human Mind*. London: Thames & Hudson.

Henrich, J. (2015). *The Secret of Our Success: How Culture Is Driving Human Evolution, Domesticating Our Species, and Making Us Smarter*. Princeton, NJ: Princeton University Press.

Hill, K. R., Wood, B. M., Baggio, J., Hurtado, A. M., & Boyd, R. T. (2014). Hunter-Gatherer Inter-Band Interaction Rates: Implications for Cumulative Culture. *PLoS ONE*, 9(7), e102806. doi:10.1371/journal.pone.0102806

Hillman, G. C., Hedges, R., Moore, A. M. T., Colledge, S., & Pettitt, P. (2001). New Evidence of Lateglacial Cereal Cultivation at Abu Hureyra on the Euphrates. *The Holocene*, 11(4), 383–93.

Hodder, I. (1990). *The Domestication of Europe: Structure and Contingency in Neolithic Societies*. Oxford: Basil Blackwell.

Hodder, I. (2011). Human-Thing Entanglement: Towards an Integrated Archaeological Perspective. *Journal of the Royal Anthropological Institute*, 17(1), 154–77. doi:10.1111/j.1467-9655.2010.01674.x

Hodder, I. (2016). *Studies in Human-Thing Entanglement* [pdf]. Retrieved from http://www.ian-hodder.com/books/studies-human-thing-entanglement

Huxley, J. (1942). *Evolution: The Modern Synthesis*. London: Allen & Unwin.

Jablonka, E., & Lamb, M. J. (2005). *Evolution in Four Dimensions: Genetic, Epigenetic, Behavioral, and Symbolic Variation in the History of Life*. Cambridge: MIT Press.

Jablonka, E., & Lamb, M. J. (2007). Précis of Evolution in Four Dimensions. *Behavioral and Brain Sciences*, 30(4), 353–65. doi:10.1017/S0140525X07002221

Kendal, J., Tehrani, J. J., & Odling-Smee, J. (2011). Human Niche Construction in Interdisciplinary Focus. *Philosophical Transactions of the Royal Society B: Biological Sciences*, 366(1566), 785–92. doi:10.1098/rstb.2010.0306

Kirby, S., Dowma, M., & Griffiths, T. L. (2007). Innateness and Culture in the Evolution of Language. *Proceedings of the National Academy of Sciences of the United States of America*, 104(12), 5241–5.

Laland, K., Matthews, B., & Feldman, M. W. (2016). An Introduction to Niche Construction Theory. *Evolutionary Ecology*, 30(2), 191–202. doi:10.1007/s10682-016-9821-z

Laland, K., & O'Brien, M. (2010). Niche Construction Theory and Archaeology. *Journal of Archaeological Method and Theory*, 17(4), 303–22. doi:10.1007/s10816-010-9096-6

Laland, K., & O'Brien, M. (2011). Cultural Niche Construction: An Introduction. *Biological Theory*, 6(3), 191–202. doi:10.1007/s13752-012-0026-6

Laland, K. N. (2017). *Darwin's Unfinished Symphony: How Culture Made the Human Mind*. Princeton, NJ: Princeton University Press.

Laland, K. N., Odling-Smee, J., & Feldman, M. W. (2001). Cultural Niche Construction and Human Evolution. *Journal of Evolutionary Biology*, 14(1), 22–33. doi:10.1046/j.1420-9101.2001.00262.x

Laland, K. N., Uller, T., Feldman, M. W., Sterelny, K., Müller, G. B., Moczek, A., . . . Odling-Smee, J. (2015). The Extended Evolutionary Synthesis: Its Structure, Assumptions and Predictions. *Proceedings of the Royal Society B: Biological Sciences*, 282(1813), 20151019. doi:10.1098/rspb.2015.1019

Liu, C., & Stout, D. (2022). Inferring Cultural Reproduction from Lithic Data: A Critical Review. *Evolutionary Anthropology: Issues, News, and Reviews*, 1–17. doi:10.1002/evan.21964

Menary, R. (2010). The Extended Mind. Cambridge: MIT Press.

Mesoudi, A., & Thornton, A. (2018). What Is Cumulative Cultural Evolution? *Proceedings of the Royal Society B: Biological Sciences*, 285(1880). doi:10.1098/rspb.2018.0712

Mithen, S. (1996). *The Prehistory of the Mind: A Search for the Origins of Art, Religion and Science*. London: Thames & Hudson.

Moore, A. M. T., & Hillman, G. C. (1992). The Pleistocene to Holocene Transition and Human Economy in Southwest Asia: The Impact of the Younger Dryas. *American Antiquity*, 57, 482–94.

Muthukrishna, M., Doebeli, M., Chudek, M., & Henrich, J. (2018). The Cultural Brain Hypothesis: How Culture Drives Brain Expansion, Sociality, and Life History. *PLOS Computational Biology*, 14(11), e1006504. doi:10.1371/journal.pcbi.1006504

Odling-Smee, F. J., Laland, K. N., & Feldman, M. W. (1996). Niche Construction. *American Naturalist*, 147(4), 641–8.

Odling-Smee, F. J., Laland, K. N., & Feldman, M. W. (2003). *Niche Construction: The Neglected Process in Evolution*. Princeton, NJ: Princeton University Press.

Powell, A., Shennan, S., & Thomas, M. G. (2009). Late Pleistocene Demography and the Appearance of Modern Human Behavior. *Science*, 324(5932), 1298–301. doi:10.1126/science.1170165

Reader, S. M., & Laland, K. N. (2002). Social Intelligence, Innovation, and Enhanced Brain Size in Primates. *Proceedings of the National Academy of Sciences of the United States of America*, 99(7), 4436–41. doi:10.1073/pnas.062041299

Rendell, L., Fogarty, L., & Laland, K. N. (2011). Runaway Cultural Niche Construction. *Philosophical Transactions of the Royal Society B: Biological Sciences*, 366(1566), 823–35. doi:10.1098/rstb.2010.0256

Sánchez-Villagra, M. R., & van Schaik, C. P. (2019). Evaluating the Self-Domestication Hypothesis of Human Evolution. *Evolutionary Anthropology: Issues, News, and Reviews*, 28(3), 133–43. doi:10.1002/evan.21777

Smith, B. D. (2016). Neo-Darwinism, Niche Construction Theory, and the Initial Domestication of Plants and Animals. *Evolutionary Ecology*, 30(2), 307–24. doi:10.1007/s10682-015-9797-0

Sterelny, K. (2011). *The Evolved Apprentice: How Evolution Made Humans Unique*. Cambridge: MIT Press.

Sterelny, K. (2019). The Archaeology of the Extended Mind. In M. Colombo, E. Irvine, & M. Stapleton (Eds.), *Andy Clark and His Critics* (pp. 143–58). New York, NY: Oxford University Press.

Stout, D. (2011). Stone Toolmaking and the Evolution of Human Culture and Cognition. *Philosophical Transactions of the Royal Society B: Biological Sciences*, 366(1567), 1050–9. doi:10.1098/rstb.2010.0369

Stout, D. (2019). Homo Artifex: An Extended Evolutionary Perspective on the Origins of the Human Mind, Brain, and Culture. In K. A. Overmann, K. A. Overmann, F. L. Coolidge, & F. L. Coolidge (Eds.), *Squeezing Minds from Stones: Cognitive Archaeology and the Evolution of the Human Mind* (pp. 42–58). Oxford: Oxford University Press.

Theofanopoulou, C., Gastaldon, S., O'Rourke, T., Samuels, B. D., Messner, A., Martins, P. T., ... Boeckx, C. (2017). Self-Domestication in Homo sapiens: Insights from Comparative Genomics. *PLoS ONE*, 12(10), e0185306. doi:10.1371/journal.pone.0185306

Tomasello, M. (2014). The Ultra-Social Animal. *European Journal of Social Psychology*, 44(3), 187–94. doi:10.1002/ejsp.2015

Turchin, P. (2013). The Puzzle of Human Ultrasociality: How Did Large-Scale Complex Societies Evolve? In P. J. Richerson & M. H. Christiansen (Eds.), *Cultural Evolution: Society, Technology, Language, and Religion* (pp. 61–73). Cambridge: MIT Press.

Watkins, T. (1990). The Origins of House and Home? *World Archaeology*, 21(3), 336–47.

Watkins, T. (2004a). Architecture and 'Theatres of Memory' in the Neolithic of South West Asia. In E. DeMarrais, C. Gosden, & C. Renfrew (Eds.), *Rethinking Materiality: The Engagement of Mind with the Material World* (pp. 97–106). Cambridge: McDonald Institute of Archaeological Research.

Watkins, T. (2004b). Building Houses, Framing Concepts, Constructing Worlds. *Paléorient*, 30(1), 5–24.

Watkins, T. (2012). Household, Community and Social Landscape: Building and Maintaining Social Memory in the Early Neolithic of Southwest Asia. In M. Furholt, M. Hinz, & D. Mischka (Eds.), *'As Time Goes By' Monuments, Landscapes and the Temporal Perspective. Socio-Environmental Dynamics over the Last 12,000 Years* (Vol. 206, pp. 23–44). Kiel; Bonn: Rudolf Habelt.

Watkins, T. (2017). Architecture and Imagery in the Early Neolithic of Southwest Asia: Framing Rituals, Stabilizing Meanings. In C. Renfrew, I. Morley, & M. Boyd (Eds.), *Ritual, Play and Belief in Early Human Societies* (pp. 129–42). Cambridge: Cambridge University Press.

Zeder, M. A. (2017). Domestication as a Model System for the Extended Evolutionary Synthesis. *Interface Focus*, 7(5), 20160133. doi:10.1098/rsfs.2016.0133

10 The Epipalaeolithic–Neolithic Transformation

The Pivot of Cultural Evolution

In the earlier chapters, I sketched out the archaeological information that describes, stage by stage, the ENT. The key question, which has been the constantly debated subject for more than seven decades, is what was it that drove this transformation; and I would add the related question of why it happened around the Pleistocene–Holocene boundary and not earlier. It has long been an inconclusive debate between two versions of an ecological pressure as the driver: on the one hand, the pressure of increasing population on finite wild food resources, or alternatively the adverse impact of the Younger Dryas climatic phase on the environment, reducing natural food resources. That this debate between rival ecological models remains inconclusive after 70 years suggests that it is time to look elsewhere.

In the previous chapter, I proposed that we are now seeing how cultural evolutionary theory and the theory of the extended mind are converging to explain the long-term emergence and evolution of our human uniqueness. The key ideas are cultural niche construction and cumulative cultural evolution: over the long-term humans have evolved a cultural niche that not only ensures the secure transmission of an increasingly complex body of cultural knowledge from generation to generation, but also allows, or even encourages, innovation and the gradual accumulation of cultural knowledge. In this chapter I want to bring the archaeology of the Epipalaeolithic–Neolithic periods together with those cultural evolutionary ideas from the last chapter. That will enable us to recognise that our period saw a radical transformation of the cultural niche that enhanced its capacity to ensure the long-term maintenance of an expanded and complex cultural package. At the same time, the rapidly developing cultural niche successfully required, but also supported, more efficient and more effective cumulative culture. I am following closely Joe Henrich's account of how 'culture is driving human evolution, domesticating our species, and making us smarter' as 'the primary driver of our species' evolution' (Henrich 2015: 57).

The ENT brought about a step-change in the scale and tempo of human social and cultural development. Against the unimaginable time-frame of the Palaeolithic, whether the hundreds of thousands of years of deep Palaeolithic, or the tens of thousands of years of Homo sapiens' early history, the scale and tempo of cultural, social and economic change across the fifteen millennia of the ENT is at a quite revolutionary new rate. What we observe within our period is a marked

DOI: 10.4324/9781351069281-11

upward kink in what had been a very slowly accelerating curve; and what came afterwards, beyond the end of the Neolithic period, continued happening on an expanding scale and tempo.

A transformation involves something that existed being changed into something different. Within southwest Asia, around the arc of the hilly flanks and in central Anatolia and Cyprus people found advantages in making new ways of living together that involved sedentism, larger, denser and more intensive networks, and new ways of managing and intensifying their food production through cultivation of crops and the herding of animals. None of these aspects of the new way of life was entirely novel: people had always lived together, and human communities had been engaged in networking their social relations for many tens or hundreds of thousands of years. Earlier Palaeolithic societies had sought to manage the resources of their environment. To take just two examples from recently published research; around 125,000 years ago, Middle Palaeolithic Neanderthal groups had adapted the natural forested environment around the shores of a lake in Germany, creating an open landscape which enabled them to occupy their preferred camp sites for longer stays (Roebroeks et al. 2021). The second example is simply the latest contribution to a long-running debate concerning the role of human hunters in the extinction of many large animal species (Ben-Dor & Barkai 2021). By the time that our period begins, the largest mammals that once roamed the Levant had disappeared or were close to extinction, and over the Upper Palaeolithic and Epipalaeolithic periods Flannery's broad-spectrum revolution can be documented in the reduction in the range of available herbivores and the increasing focus on gazelle, and other smaller prey, including birds, reptiles and fish. What is impressive about the ENT is the transformation of old ways, the relative rapidity of that transformation relative to earlier Palaeolithic times, the scale of the cultural, social and economic transformation, and the foundations that it laid down for further diverse cultural, social, technical and economic change.

The difference between the archaeology that we have from the Upper Palaeolithic and what we see in the established Neolithic is massive. In earlier chapters we observed the changing panorama through the millennia of the Epipalaeolithic and onwards with increasing rapidity and scale through the Pre-Pottery Neolithic, and then through further critical changes into the Pottery Neolithic. One reason why the transformation took place within certain parts of southwest Asia is that the hilly flanks of the Fertile Crescent plus parts of central Anatolia were extraordinarily rich in both plant and animal resources for hunter-gatherers, including storable resources such as seeds and nuts. And among those resources was the broadest range of potential domesticates. Of all the zones around the world where domestications occurred, those key parts of southwest Asia are the best endowed to support the ENT. But the availability of potential domesticates is not enough.

For all but the last 200,000 years or so, the story of human evolution has been about the co-evolution of the biological brain, its cognitive capacity to manage increasing numbers of increasingly complex social relations, and the scale and nature of human societies. The main story-line that we get from the Lucy to Language research project (Dunbar et al. 2014; Gamble et al. 2014) reaches its end when it comes to Homo sapiens. But with Homo sapiens the evolutionary story has to take a different turn. Homo sapiens has been around, to the best of our present knowledge, for about 300,000 years. In terms of physical biological evolution of the size or structure of the brain, Homo sapiens has not been in existence long enough to have evolved; despite all the variety of Homo sapiens population around the world, there is next to no variation in the human genome. But there is a world of difference between us today and the technology and culture by which we live and the technical and cultural world of Homo sapiens in the Upper Palaeolithic, and another world of difference between Homo sapiens of 30–40,000 years ago and what we have been learning from sites occupied by Homo sapiens in southern Africa of 100,000 years ago (not to mention the formalised burials of Homo sapiens bodies accompanied by red ochre in caves in northern Israel dated as early as 120,000 years ago). In short, the massive changes in human society and culture of the last 100,000 years (including those that are the subject of this book of 23,000–8000 years ago), representing the most dramatic chapters in the whole of human evolution, must be understood in terms of the ways that humans have evolved and continue to evolve new cultural and cognitive ways of working in ever more complex and powerful modes.

The transformation, therefore, was brought about by cultural means. While there has been continual gene-culture co-evolution, the story of the ENT, which is relatively brief in biological evolutionary terms, does not depend on further biological evolution of the human brain: multiple human minds devised new ways (cumulative cultural evolutionary ways) to live together permanently, thereby transforming their cultural niche.

Long-Term Cultural Evolution: A Story of Accelerating Scale and Tempo

My purpose is to show that what was happening in the ENT was a continuation of earlier Pleistocene–Palaeolithic cultural evolution, but with a transformative change in tempo and scale. As we saw in the previous chapter, human cultural evolution has relied on the constant creation and recreation of the cultural niche; and the cultural niche has evolved to ensure the surest means of sustaining over time an increasingly complex cultural package. For most of human cultural evolution that capacity for cultural accumulation has been scarcely perceptible in the archaeological record except in the occasional developments and shifts in stone tool-making modes. Since the appearance of Homo sapiens cultural accumulation has become more apparent. From at least 120,000 years ago it became significant in more ways than improving and diversifying stone tool technology. The human cultural niche has provided for not only the safe retention of the group's cultural knowledge, skills and traditions but also for cultural cumulation – the capacity to generate and incorporate innovations. That capacity for cumulative culture, it is argued in this chapter, goes up a gear in scale and tempo and is central to the ENT.

Over and above the morphological changes in the body and the brain in hominin evolution over the last two or three million years, there are three significant, inter-locking trends in human social and cultural evolution. Once we have taken a quick look at those long-term trends, we can focus on the ENT, where those three inter-locking trends can be identified in operation, but at a rapidly accelerating rate. Those three inter-related, slow, but accelerating trends are documented in the archaeological record: (a) in cultural innovation and change; (b) in the expansion of the range of cultural products, skills, and capacities; (c) and in the growth of population, of population density and the scale of human social groups. The first two of those trends are products of cumulative cultural evolution, while the third represents a fundamental transformation of the human cultural niche.

The Emergence of a New Kind of Cultural Niche

It has taken a long time to assemble the archaeological information and put it in order; it has also taken time to develop new ways of analysing and investigating different kinds of archaeological material. There is now a mass of material, although it is not evenly distributed and there are still parts of the region which are still terra incognita, where there have been far too few archaeological sites found and investigated. In this chapter, we need to extract some generalisations out of the detail, and build some understanding of the transformation, setting it as a significant episode within the cultural evolutionary framework discussed in the previous chapter. I will divide the archaeological ENT into a story in four stages.

The first phase covers most of the Epipalaeolithic period, as we see it in the Levant, where it is best documented at present. By contrast with the Upper Palaeolithic period, the number of sites in the Epipalaeolithic increases, the degree of mobility of forager groups reduces; at sites like Ohalo II there are signs that a group of hunter-gatherers stayed seasonally at an ecotone location from which a wide range of different food resources were on hand. They harvested wild cereals and other grass seeds, and presumably stayed in their cluster of huts while that harvest was consumed. Later in this first period, in seasonal wetland areas within the semi-arid of north Jordan, several aggregation sites have been identified, where numbers of hunter-gatherer groups gathered in seasons of plenty. Epipalaeolithic groups exchange mobility for what Kent Flannery called the broad-spectrum revolution, which involves greater investment in hunting and trapping small animals, birds, and reptiles, fish and shellfish, plus increasing investment in the harvesting of storable seeds, grains, and nuts, in return for seasonal stays at prime locations. There are open-air sites where permanent structures were built. Although they may not have been inhabited throughout the year, they were certainly occupied often and repeatedly, so that archaeological deposits were accumulated.

The second phase includes the last part of the Epipalaeolithic and the early Pre-Pottery Neolithic periods. The division of Pleistocene from Holocene, and of Epipalaeolithic to Neolithic at 9600 BC is fixed. Over the end of the Epipalaeolithic and the early Pre-Pottery Neolithic we see populations become fully sedentary, living in what seem to us to be small settlements that nevertheless, by contrast with earlier periods, represent people living together in larger numbers in permanently co-resident societies. For a long time, archaeologists have tried to discover whether the beginning of early Pre-Pottery Neolithic settlements was triggered by the onset of the Holocene climatic optimum. On the one hand, the impact of the Younger Dryas cold phase in southwest Asia has proved elusive and diverse, which makes it difficult to contrast the beginning of the Holocene with the preceding period; and, on the other hand, we have been learning of a string of settlements in the Tigris valley in southeast Turkey that on the basis of radiocarbon dates were founded in the middle of the Younger Dryas period, and show no change as they progress into the early Pre-Pottery Neolithic period. There is good reason to put together into a single phase the later Epipalaeolithic and the early Pre-Pottery Neolithic.

How to understand late Epipalaeolithic sites such as Eynan, where there is a stratigraphic succession of permanent buildings that in total cover more than 2000 years, remains to be resolved: is it possible that what looks like a permanent settlement of successively rebuilt stone houses was continuously occupied throughout that time? Even if the site was occupied and reoccupied with breaks between one occupation and the next, such continuity of concern for the location is remarkable. Whether in the late Epipalaeolithic or the early Pre-Pottery Neolithic, the sedentary way of life in a permanent settlement depended on hunting and gathering within the territory around their settlements, plus they managed crops of both cereals (primarily wheat and barley) and legumes (notably pulses such as lentils and chickpeas).

Coming together to live in permanent settlements required the greater implementation of ritual activities and the construction of large and elaborate communal buildings (although we are frustrated when we want to know how those buildings were used). The burial of their dead in places where they had lived was a practice that had begun as far back as the late Middle Palaeolithic. In the early Epipalaeolithic there have been burials found among the structures of open-air settlement sites; and in the later Epipalaeolithic there are sites with clusters of elaborate burials under structures, or small cemeteries of burials associated with a rock-shelter occupation site. In the early Pre-Pottery Neolithic there is a range of burial practice within the permanent settlements. At Jerf el Ahmar beside the Euphrates in the north of Syria, there was a succession of large, elaborate circular subterranean communal buildings, but no burials within the settlement. Along the Tigris valley in southeast Turkey, each settlement was different in terms of the numbers of intramural burials, with Körtiktepe topping the table with several hundred bodies buried below the floors of the houses. The new, larger-scale societies needed new or enhanced social mechanisms to ensure social cohesion; Sterelny (2018, 2020) shows how the emergence of what he calls 'articulated religion' involved the costly signalling of collective rituals within the linked stories of a mythology or ideology. In the new larger and more complex context of super-communities made up of (relatively) large, sedentary communities, that articulation took the form of architectural symbolism, sculpture and iconic symbols. From another perspective, the new cultural niche required new social institutions to defuse the increased stresses of living in sedentary communities and dampen the inevitable conflicts; Dunbar (2022) argues that the community-level institutions and rituals such as the creation and maintenance of communal buildings (and their use) served to enhance the sense of belonging to community bonding.

The third phase covers most of the later Pre-Pottery Neolithic period. The size of co-resident communities grew, and, in a number of cases, there would have been many hundreds or several thousand people living together in settlements that persisted for many centuries. In this phase the subsistence economy was increasingly dependent on domesticated cereals and pulses and the herding of sheep and goat. Wild boar would have been around in many parts of our region, but they were not hunted for whatever reason by many Palaeolithic hunter-gatherer groups; similarly, although pig would have been fairly readily domesticated, they are not found at many later Pre-Pottery Neolithic settlements. Wild cattle were also only taken under human control and domesticated in a few places; cows for milking and oxen for ploughing become significant in the fourth phase.

Archaeologists used to talk of an extensive PPNB culture, or, when the culture concept became questionable, a PPNB interaction sphere (Bar-Yosef & Belfer-Cohen 1989). It is certainly true that social and cultural networking in the later Pre-Pottery Neolithic was both more intensive and more extensive. One dimension of the complexity and intensity of networking is illustrated by the work of the Spanish group who have analysed the distribution of central Anatolian obsidian throughout settlements in the Levant. What is striking is the range of sizes of settlements at this period, both throughout the length of the Levant, and beyond. There are many small settlements, as there were in the earlier Pre-Pottery Neolithic, but in the later Pre-Pottery Neolithic we see the emergence of a number of larger settlements to the extent that the largest of them have been labelled 'mega-sites'.

At one level, each of these permanent communities needed to attend to the needs of social cohesion and social bonding, resulting in the individual characteristics of each of them in the archaeological record: the perception that at the same time there were very extensive 'interaction spheres' within which these societies were able to share and engage is important. Rather than the PPNB interaction sphere recommended by Bar-Yosef and Belfer-Cohen, I prefer the peer-polity interaction sphere model that Colin Renfrew (1986) proposed: that kind of interaction sphere where all the participating communities show by their sharing and exchange that they share the same values. While demonstrating their shared community, they also recognise that each has its own way of doing so; a simple analogy would be the way that football clubs compete within the rules of the game, but each has its own distinctive uniform of shirts and shorts. Renfrew emphasised the importance of competitive emulation within the interaction sphere. Inter-communal competition leading to conflict has been seen to be a risk, and archaeological examples of warfare have been found, for example in Neolithic Britain; and inter-community warfare is not uncommonly encountered in the ethnographic literature. Competitive emulation involves competing within the interaction sphere by emulating others in order to maintain or enhance status within the group. That can be done by means of the strength of the community's social exchange relations, and the social exchanges may be accompanied by the exchange of certain goods or materials. The things in the social exchange networks, such as obsidian or marine shells for our Epipalaeolithic–Neolithic, had become standardised, a process that Renfrew calls 'symbolic entrainment'. In Renfrew's view, which was developed out of his research on the Aegean Early Bronze Age, the peer-polity interaction sphere enhanced the 'transmission of innovation' and minimised the risk of inter-communal conflict. There is much in Renfrew's model of the peer-polity interaction sphere, I believe, that helps us to perceive the working of the new cultural niche of sedentary communities locked into super-communities. We may think of the super-communities of the Epipalaeolithic–Neolithic transformation as the invention of Renfrew's peer-polity interaction sphere.

The fourth phase starts with the end of the Pre-Pottery Neolithic and continues through what archaeologists consequently call the Pottery Neolithic. Archaeologists have generally concentrated on the Pre-Pottery Neolithic, which has produced some spectacular (size and architecture) and extraordinary (intramural burial and modelled skulls) archaeology. By contrast, the fourth phase of the process saw the abandonment of almost all of the large, classic later Pre-Pottery Neolithic settlements, and the spread of many, smaller, less densely built-up settlements across a much wider area than that of the previous phases. Settlement within southwest Asia spread out beyond the hilly flanks, implementing new adaptations as it extended into the drier tracts of inland Syria and Jordan, new subsistence economic strategies across the green Jezirah of north Mesopotamia, and new irrigation technology in the alluvial lands of southern Iraq and southwest Iran. And, to mirror the expansion within the curve of the Fertile Crescent, there was an outward expansion, whose outline we shall trace only in a westward direction, from northwest Anatolia into the Balkans, into the western Anatolian coastlands, the Aegean islands, and the Greek mainland – the beginning of an extraordinarily rapid expansion across Europe.

Accelerating Scale and Tempo within the ENT

For the most part, studies of cultural evolution have concluded either with the emergence of Homo sapiens, or around the beginning of the Upper Palaeolithic in European terms. Perhaps too many were impressed by the enthusiasm with which influential Palaeolithic archaeologists embraced the idea of a 'human revolution' with the arrival of Upper Palaeolithic art and symbolism (Mellars & Stringer 1989). That was proclaimed as the emergence of the modern mind, and the rest was history. To Colin Renfrew it was a curious omission in view of the extraordinary amount of social and cultural evolution of the last 50,000 years; he called it 'the sapient paradox' (Renfrew 1996, 2008). From the opposite direction, the idea of the human revolution and the emergence of 'modernity' was undermined by McBrearty and Brooks (2000), pointing out the evidence from southern Africa of the precursors of Upper Palaeolithic symbolism.

Now I want to argue that we can see those same three characteristics of long-term hominin evolution (cultural innovation and change, the expansion of the range of cultural products, skills, and capacities, and the growth of population, of population density and the scale of human social groups) operating within our ENT, but at new rates of acceleration and scale that are of a different order when seen against the unimaginably long term of hominin evolution. And this relatively sudden and dramatic acceleration sets the scene for all that follows in human cultural evolution.

The accelerating rate of cultural change is implicit in the way that archaeologists have defined successive archaeological periods in terms of their changing material culture, which in effect has meant changes in chipped stone tool-making methods and products. Tracing cultural change within the tens of thousands of years of the later Middle Palaeolithic has proved extremely difficult. The Middle Palaeolithic chipped stone assemblages have been subjected to many careful studies, especially since it was recognised that there were both Homo sapiens and Neanderthal populations in the Levant from around 120,000 years ago. We do not have to engage with the conundrum of why it is practically impossible to distinguish different technological traditions; for us it is enough to note that there is no definable change in chipped stone tool-making that allows the specialists to recognise the arrival of Homo sapiens, and nothing to distinguish the chipped stone tools of around 50,000 years ago from those of 120,000 years ago. The rate of cultural evolution, whether in the hands of Neanderthals or Homo sapiens, was imperceptible over those tens of thousands of years. When we cross the boundary between the Middle and Upper Palaeolithic, things begin to change. For a start, the Neanderthals have disappeared; and, at least in the Levant, there are two distinct chipped stone traditions within that period of around 25,000 years. The Epipalaeolithic period is half the length of the Upper Palaeolithic, and in the Levant, where there has been most research over the last hundred years, the Epipalaeolithic is broken down into three sub-periods on the basis of changes in the ways the microlithic tools were made. The period specialists would also insist that there were regional variants within the Levant in any of those sub-periods. The phases within the Pre-Pottery Neolithic are shorter again. The early Pre-Pottery Neolithic, often called the PPNA, lasts about 2000 years, and some specialists identify an initial Khiamian phase lasting a mere few hundred years. As more sites in the Levant have been investigated and more research has been devoted to the period, it is becoming possible to differentiate an early PPNA from a late PPNA. The later Pre-Pottery Neolithic, often labelled PPNB, lasted for about 2000 years, and consists of four sub-periods. Where we count in tens of thousands of years at the beginning of the Upper Palaeolithic sequence, we count in a few centuries for each sub-period towards the end of the Neolithic.

The material culture repertoire can equally be seen to expand over our period, in particular through the Pre-Pottery Neolithic. Going back to Gordon Childe's view of the Neolithic, and even further to the original formulation of a Palaeolithic and a Neolithic in the nineteenth century, the Neolithic was marked by the addition of new craft skills, including polished stone tools, weaving, and farming. New kinds of artefacts document the new skills that were being added. A modern take on the expansion of material culture has been developed by Ian Hodder. His essays and book on how humans are 'entangled' with 'things' have appeared over the latter years of Ian's 25 year entanglement with research at Çatalhöyük (Hodder 2011a, 2011b, 2012, 2020). Hodder's 'things' are not confined to artefacts, but 'to naturally occurring objects, animals, plants, and humans, as well as sounds and words – any object or sound in which humans have an interest' (Hodder 2011a: 155).

He notes that, by contrast with earlier periods, across the Neolithic period, with which he was mainly concerned, 'the amount of stuff in peoples' lives increased dramatically' (Hodder 2014: 28).

Across our Epipalaeolithic–Neolithic period, the first simple buildings made of stone or mud-brick and mortar became multi-roomed and sometimes two-storey multi-roomed buildings. In addition to their deep inheritance of wild plants, they acquired farmed cereals and legumes; digging, sowing, weeding, reaping, thresh-ing, winnowing were all new skills, and they required new kinds of tools. While they might still continue to hunt, in the Neolithic they began to manage flocks of sheep and goat, and maybe some pigs and cattle. Knowing how to thresh and win-now a harvest, how to prepare grain for storage, and how much to reserve for next season's sowing was of crucial importance, as was knowing how to fodder your animals when grazing was not available, how many animals to carry through the lean season of winter, and how to manage the breeding cycle of your animals. All of this represented the acquisition of new knowledge and a host of new skills.

During the Neolithic, people added pottery made of clay and fired in ovens; pottery of all kinds served new ways of processing, storing, cooking and serving food. People knew how to get fibre from plants, but in the Neolithic they learnt how to spin and weave textiles made from wool. There are certainly more beads and personal ornaments in greater variety than previously (Bar-Yosef Mayer 2013). Among them there are beads made from malachite, an oxide of copper, and galena, or lead, at a number of sites from southwest Iran (Ali Kosh) to central Anatolia (Aşıklı Höyük and Çatalhöyük). In some cases these early copper objects were present in quantities and show a degree of metallurgical skill. More than 100 small copper objects have been recorded at later Pre-Pottery Neolithic Çayönü, includ-ing beads, hooks, awls, and pins. They are made of native copper (that is, from naturally occurring surface nuggets), but some of them at least have been shown to have been formed by hot-working (Özdogan & Özdogan 1999). More than 100 burials were found below house floors at Tell Halula, a late Pre-Pottery Neolithic settlement in the Euphrates valley in north Syria, and more than half of the bodies were buried with personal ornaments, such as stone beads and shells that were part of necklaces, bracelets and head- dresses (Kuijt et al. 2011). Eleven of those burials have produced copper beads and a copper lunula; detailed examination has shown that they were made by repeated annealing and cold-hammering.

So we can recognise that there was an acceleration in the rate and the range of cultural cumulation, whether artefacts, skills, knowledge, or concepts. The theo-ries of cultural niche construction and cumulative culture would expect that such acceleration in the rate and expansion in the range of cultural growth should be accompanied by – essentially supported by – larger, more intensively cohesive populations.

We can get a proxy handle on the growth of population and population density in our period by means of occupation sites and settlements. Estimating actual num-bers of individuals, whether of the group using a rock-shelter or the community liv-ing together in a permanent settlement, is next to impossible. We can only manage this subject by taking the gross counts of numbers of sites or sizes of settlements per archaeological period. Nigel Goring-Morris and Anna Belfer-Cohen (2011) brought together the data on the number of sites in different parts of southwest Asia between the beginning of the Upper Palaeolithic (around 50,000 years ago) and the late Neolithic (around 8000 years ago). For the purposes of graphing the data (p. S199, Figure 2), the numbers of sites were normalised relative to the duration of each sub-period. For the southern Levant, where the best data has been accumulated from more than a century of fieldwork, the number of sites grows steadily from period to period in a roughly straight line. Over the last 50 years there has been a concentration of salvage archaeology on the upper Euphrates and Tigris rivers both in north Syria and in southeast Turkey. Although we do not yet have early (Upper Palaeolithic and Epipalaeolithic) sites in that region, Goring-Morris and Belfer-Cohen's graph shows a similar straight line representing the growth in the numbers of sites from the beginning of the Neolithic onwards. Both straight lines in fact under-play actual population growth, because across time (a) sites became larger, (b) (at least in the southern Levant) they were occupied more permanently, and (c) our archaeological periods through the Epipalaeolithic and the Neolithic get shorter and shorter with time. If we redrew the graph for the southern Levant with the horizontal scale scaled for time, we would produce an accelerating upward curve.

Ian Kuijt (2000) collected data on Neolithic settlement size for the southern Levant. His graph (p. 83, Figure 2) shows that average site size increased across the Pre-Pottery Neolithic in an accelerating upward curve. In another diagram (p. 90, Figure 6) he graphed the ratio of built space to open space (how scattered or densely packed the buildings were). As settlements grew in size through time, so did the density of buildings within them. In the late Epipalaeolithic site of Eynan there was more open space compared to roofed space, while in settlements of the late Pre-Pottery Neolithic there was four to eight times more roofed than open space between the buildings. These two accelerating curves together amplify even further the dramatic crescendo of growth in settlement size, density of housing and population. In settlements of the end of the Epipalaeolithic and the earliest Pre-Pottery Neolithic there may have been one or two hundred people; estimates of population in the largest settlements of the later Pre-Pottery Neolithic range in the thousands. Detailed study of the massive settlement of Çatalhöyük, for example (Cessford 2005; Hodder 2020: 74), has led to an estimate of population at its height of between 3500 and 8000. The uncertainty of those figures is caused by the difficulty of identifying precisely how many houses were in active use and how many were abandoned and treated as rubbish dumps at any one time.

An important feature of successful and resilient societies is their cohesiveness. Whatever the means by which they assured their internal social cohesion, our Neolithic communities did not exist in isolation. For a long time the close similarities in material culture among the sites across a region were taken to be the common features that define an archaeological culture in the sense defined almost a century ago by Gordon Childe. But over the decades we have learned of many more settlements, and we have much more detailed information; it has become clear that each Neolithic settlement resembled its neighbours in general terms, but was distinct and idiosyncratic in subtle ways. The sharing of cultural practices among a number of culturally autonomous communities must be a token of the maintenance of close social and cultural relations.

We can document the negotiation of relationships of sharing and exchange through the exchange of materials and things. The analyses of the distribution central and east Anatolian obsidian were discussed in Chapter 8. First Colin Renfrew and his collaborators (1968, 1976) showed that, within a radius of 250–350 km of the central Anatolian sources, communities such as Çatalhöyük and Aşıklı Höyük relied on obsidian for their chipped stone tools; Renfrew supposed that they supplied themselves with an essential raw material, and he called that the supply zone. Beyond the supply zone amounts of obsidian dropped dramatically to 10% or less of the total chipped stone. Renfrew called this outer zone the contact zone; the amounts at Neolithic settlements declined with distance, from 10% to 1% and to 0.1% in southern Jordan. Now we know a good deal more about other materials and artefacts that were exchanged. We also know that the extensive networks were already in existence in the Epipalaeolithic period, and the steady growth in the amount of obsidian and the range of other materials in the network can be charted. As we saw in Chapter 8 from the work of a group of Spanish researchers (Ibáñez et al. 2016; Ortega et al. 2014), in the later Pre-Pottery Neolithic period it is necessary to suppose the operation of 'small-world networks' in which communities worked out the best means to support their social connections, which could mean that they bypassed their nearest neighbours and accessed preferred 'distant links' directly, exchanging with partners up to 180 km from home. In the later Pre-Pottery Neolithic, when volumes of obsidian were at their greatest, 'optimised distant link' networking best explains the distribution. And the modelling shows the emergence of some settlements as significant distribution centres that could obtain their obsidian direct from other centres that were nearer the Anatolian sources. In other words, the research suggests that there came into existence in the early Neolithic complex and hierarchical systems of interaction and exchange of symbolically important materials such as obsidian, marine shells, and stone beads, genes (through exchange of marriage partners), and the sharing of ideas, innovations and experiences.

It is easy to think of our individual settlement sites as autonomous communities, but we should be thinking in terms of super-communities made up of networked settlements (Watkins 2008). In addition to the proxy evidence of generally increasing population density (more and more settlements), and increasing numbers of people living permanently together (larger and larger co-resident communities), the true measure of the scale of the social group is the regional or supra-regional super-community. Across our period, and particularly across the later Pre-Pottery Neolithic, the scale and the intensity of local, regional, and supra-regional exchange and interaction increased dramatically. At the beginning of the later Pre-Pottery Neolithic, as we saw in Chapter 8, a new way of forming and extracting blades from flint or obsidian cores, the so-called naviform core technology, emerged within a cluster of settlements in the north Levant, around the present Syrian-Turkish border. Many of the cores and their products were obsidian, and their source has been traced to specialised workshops on the flanks of Göllü Dag in central Anatolia. The technique required great expertise, but it was the interest in the new technology and the high level of skill to execute it rather than the novel product itself that was rapidly transmitted south throughout the Levant. The whole picture suggests the growth of 'professional' flint and obsidian knappers operating in a network extending from central Anatolia workshops, throughout the Levant and as far as northern Saudi Arabia (Balkan-Atli & Binder 2012; Barzilai & Goring-Morris 2013; Quintero & Wilke 1995).

A New Kind of Cultural Niche

Archaeologists tend to think in terms of the settlement, or more accurately in terms of the site whose stratigraphy and buildings they are focused on understanding through excavation. We can take a broader perspective and think of the cultural niche as operating at three, hierarchical levels. The great deal of labour and care that was invested in houses and other buildings suggests that we should pay a good deal of attention to the house and the household. The houses and other buildings formed the settlements represent for us the palimpsest of a community, hundreds or thousands of people living together, dependent in complex ways on one another. Clearly there must have been some kind of social organisation within the large communities of settlements like Çatalhöyük; did groups of households function as active neighbourhoods, or did households maintain close family relations, wherever their houses were? Analyses of the human remains from the burials under the floors of the houses had shown that those buried in one house were not genetically related as family members (Pilloud & Larsen 2011). Ian Hodder and his colleagues have spent a great deal of effort trying to identify the relations between households within the dense mass of houses. There were a few houses where there were many burials (one house that had lasted for around 50 years had more than 60 bodies under its floor), while most houses had few or none; Hodder proposed that these might be 'history houses', that is, houses which were repeatedly rebuilt and which had developed specific symbolic or political or economic eminence (Hodder & Pels 2010: 164). In short, even at the most intensively and extensively researched settlement it has proved practically impossible to identify the internal social organisation of the community. And, as I have said earlier, no settlement could exist in isolation: communities were engaged in networks of interaction and inter-dependence at the local, regional and supra-regional level. I think of this level as the super-community.

Over the last 30 years we have been learning about the earliest permanent settlements of the end of the Epipalaeolithic and the early Pre-Pottery Neolithic, as we saw in Chapters 5 and 6. It is easy to recognise the novel features, the investment of effort and care that was invested in the building of small but substantial houses. And, when we look at the large, usually semi-subterranean, communal buildings and sculptured imagery, we can be amazed at the thought of the design, logistics, skills and collaborative labour that were required. At the same time, it seems that these new permanently co-resident communities retained important cultural and social ideals and practices that were central to the mobile forager way of life, the ethology of sharing. Those studying the early stages of human evolution have pointed to the way that small-scale hunter-gatherer societies collaborate in hunting, food preparation, and food sharing in ways that would distinguish our early human ancestors from their closest primate cousins. Anthropological fieldwork has documented the close kin family groups within sharing clusters of 'households' within the residential camp (e.g. Dyble et al. 2016). At Jerf el Ahmar in the early Pre-Pottery Neolithic we saw that there were small, simple houses (with no provision of a cooking hearth) clustered around large outdoor hearths. The early communal buildings at Jerf el Ahmar incorporated large-scale bins used for the storage of harvested grain and lentils. In Jordan, buildings have been identified as communal facilities with raised floors to ensure dry conditions for storage of harvests (Finlayson et al. 2011; Kuijt and Finlayson 2009); at WF16 there were small circular buildings interpreted as storage facilities immediately outside the huge circular communal building.

It seems obvious that, compared to life in mobile foraging bands, where discord between members could be dissolved if one person or a family left to join another band of the same society, life in a permanent settlement would be more stressful. Based on the area of settlements and the approximate density of houses, Ian Kuijt (2000) estimated that early Pre-Pottery Neolithic communities in the southern Levant may have numbered around 300 inhabitants. That is already six to ten times the average size of ethnographically documented mobile forager bands. It is impossible to discover the degree of inter-personal conflict that might have escalated beyond argument to fighting, to serious injury, or even death, among Upper Palaeolithic hunter-gatherer groups because there are insufficient numbers of buried bodies.

The ethnographic record of historically modern small-scale foraging societies shows that inter-personal conflicts lead to mortal injuries, the ultimate index of stress, in significant numbers; Dunbar (2022) has collected the figures from the literature, and joins the anthropologists who have shown that feasting, singing and dancing work to reinforce or enhance social and community bonding. Among ethnographically documented societies living together in somewhat larger numbers in permanent settlements Dunbar found that cross-cutting institutions such as sodalities or men's clubs served to manage the behaviour in particular of volatile young, and strict regulations governing marital arrangements. Among his sample of sedentary societies living together in more than a few hundred people Dunbar found that the rate of deaths through inter-personal conflict did not rise in proportion to the increased numbers of the group, and he noted a significant step-increase in social institutions associated with formal religions, together with 'professional' priests and buildings dedicated to ritual and religious purposes. (I am studiously avoiding using the word 'temple', which has been loosely and too enthusiastically applied to sites like the iconic Göbekli Tepe – see the next chapter, which tries to set our period within theories of the evolution of religion). Dunbar's purpose was to take the ethnographically documented correlation between scale of co-resident group, stress and mechanisms for reducing stress and enhancing social cohesion, and test how well it might model the evolution of mechanisms countering social stress through the ENT in southwest Asia. He concluded that the proportion of burials that represent individuals was not greater. His informed analyses of stress leading to inter-personal conflicts and deaths across societies that are of the same scale as our Epipalaeolithic and Neolithic societies is valuable; but I would apply his conclusions differently given the archaeological evidence in the following paragraphs.

The late Epipalaeolithic and early Pre-Pottery Neolithic communities would seem to have continued the hunter-gatherer ethology of food sharing, although they were harvesting and storing cereals and legumes. Sharing the meat and the leaves, herbs, berries, roots or tubers that have been brought in through the day's hunting and collecting is one thing, but, when the community concentrates on harvesting and storing substantial quantities of dry seeds, the fair sharing of the stored food resources presumably needed community norms of behaviour and perhaps institutions that managed allocations. As settlements became larger, and as people came to rely more and more on the cultivated crops and their flocks and herds, it would have become impossible to maintain the practice of communal sharing of food production and storage that was carried over from the Epipalaeolithic into the beginning of the Neolithic. The whole of the Neolithic was a time of dynamic change throughout, and those changes occurred at an increasing tempo. Ian Hodder sees the transformation within the late Pre-Pottery Neolithic and late (Pottery) Neolithic as 'a dialectic tension between house and community' (Hodder 2006: 396).

The acceleration in the expansion of settlements reached a climax in the latter part of the late Pre-Pottery Neolithic. Archaeologists defined the end of the Pre-Pottery Neolithic and the beginning of the rest of the Neolithic in terms of the appearance throughout the region of pottery, and the use of pottery for cooking and especially for serving food represents new social customs that swept the region. But there was much more to the transition to the late Neolithic than the appearance of pottery: as we saw in Chapter 8, it was the end of the large, dense settlements, and the spread of many smaller settlements in all directions, both within the arc of the hilly flanks of the Fertile Crescent and in every direction south, west, north and east. Following Dunbar's research on the relationship between the stresses of living in large, permanent communities and the institutions and social practices that served to contain the stresses and promote social cohesion, smaller settlements would be easier places to live, and needed less complicated and burdensome social rituals and institutions.

By the same token, if the capacity for sustaining and extending increasingly diverse and complex cultural packages, the networking of settled communities would be of even greater importance. As we saw, the material traces of networking, best known from the work on the distribution of Anatolian obsidian, was already extensive in the Epipalaeolithic period. Whereas mobile foragers had begun to use aggregation sites, when permanently settled communities became the norm from the end of the Epipalaeolithic and the start of the Pre-Pottery Neolithic, social networking would have required the more formal social relations between communities if it were to be effective. The research on obsidian distribution has shown that social exchange networks became more intensive, reaching a climax in the late Pre-Pottery Neolithic, but there is far less data from the late Neolithic. We should not interpret the absence of evidence of obsidian as evidence of absence. What we can see in the late Neolithic period are different kinds of new products, especially ceramics, new technologies, and new social practices that seem to begin at practically the same time with closely similar forms and surface decoration across extensive geographical ranges. That implies, I believe, that the socio-cultural networking was even more effective through different media in its support of extensive regional super-communities of smaller, more manageable communities.

There is much more that could be said. The formal end of the archaeological Neolithic period in southwest Asia is an arbitrary point in a story that continues to grow in pace and exciting cultural and social innovations, whether one continues to focus on southwest Asia, or to follow developments across Europe, northeast Africa, inner Asia or elsewhere. But the formal closing of the Neolithic is a good place to end this book, because it enables me to conclude on the importance of the ENT (and its slightly younger cousins in other parts of the world) within the long history of human socio-cultural evolution.

What I hope that I have shown is that the dynamic developments of the Epipalaeolithic–Neolithic period in southwest Asia represent the evolution of a new kind of cultural niche. At this point, I ask the reader to circle back to the previous chapter. There I introduced the cultural evolutionary theories of three leading inter-disciplinary researchers who each started from different disciplinary beginnings. Sterelny in his second book (Sterelny 2011) and now in his third (Sterelny 2021) has employed cultural niche construction theory, insisting on the importance of modes of 'cultural learning' and 'cumulative culture' from early in the story of human evolution. He writes of 'the Pleistocene social contract' that sustained practical egalitarian societies, and 'scaffolded' learning environments, where juveniles learn through structured trial-and-error from senior recognised expert practitioners. He recognises that the advent of settled communities was the critically important element in the Pleistocene–Holocene transition (our ENT). He writes that 'the scale of social life increased, and social worlds became less equal' (Sterelny 2021: 124); but he is mainly concerned with the nexus of relations that led from what he calls 'storage foraging' (in the Epipalaeolithic as far as we are concerned) to mixed farming (which, as we have seen, was beginning to develop from horticulture and herding at the end of the Neolithic), leading to property ownership, accumulation of wealth, inheritance and the inheritance of wealth, and degrees of instituted social inequality (which are all beyond the chronological scope of this book). This book has focused on a relatively brief period at the end of the Pleistocene social contract, showing how aggregation of population in networks of permanently settled communities transformed the cultural niche and accelerated the pace of cultural cumulation.

Joseph Henrich argues, as we saw, like Sterelny, for the power of the human cultural niche for facilitating the safe inter-generational transfer of complex knowledge and diverse skills (Henrich 2015). As promised in the subtitle of that book, Henrich showed 'How Culture Is Driving Human Evolution'. He showed how, from an early stage in the human evolutionary story, 'cultural evolution became the primary driver of our species' genetic evolution' (Henrich 2015: 57), becoming an accelerating process of interactions that Henrich repeatedly describes as 'autocatalytic', meaning that it produces the fuel that propels itself. In short, as was noted in Chapter 9, it becomes a runaway process, a term that is also used by Kevin Laland (Rendell et al. 2011). We also saw in Henrich's work how population size and the cohesiveness of communities was critically important. The ENT was founded on the basis of more powerful networks of larger, permanently co-resident communities that invested great importance in their social and cultural cohesion. Many of them continued for many centuries, and the general trend, at least until the end of the Pre-Pottery Neolithic period, was to increasing scale of population.

The biologist Kevin Laland was the third authority on whom I have relied in Chapter 9. His approach has been through gene-culture co-evolution and its effects on the evolution of human culture and the human mind. His early research was directed to the development of niche construction theory, and his work on the idea of cultural niche construction theory has been of great importance. Like Henrich, he has used the terms 'autocatalytic' and 'runaway' to describe the complex processes whereby a broad array of feedback mechanisms have interacted to accelerate human cognition and culture 'in a runaway, autocatalytic process' (Laland 2017: 323).

The thesis that I have been developing is that the ENT represents the evolution of a new kind of cultural niche: the ability to form large, permanently co-resident, resilient communities increased their capacity to sustain an increasingly complex package of cultural knowledge and practices; and the maintenance of extensive super-communities of sharing and exchange further enhanced the capacity of the cultural niche to sustain very complex packages of knowledge, skills and ideas. The rate of cultural cumulation began to take off. In addition, the new cultural niche was successful in biological terms supporting an increasing rate of overall population growth, and successful in cultural terms in that it supported an increased rate of cultural cumulation. Within the long term of human cultural evolution, the ENT represents a distinct upkick in the graph of population growth, and in the rate of cultural expansion, in short, in the effective implementation of that 'runaway, autocatalytic process'. Following the end of the Neolithic, the rate and the diversity of innovation continued to grow and grow. The ENT was, indeed, the formation of the platform on which has been built much of the rest of history, to our own times.

After a brief excursion into the problem of how to understand religion in the context of the Neolithic, and where our ENT sits in the cultural evolution of religion, I will return to the subject of aggregation and the acceleration of expansion and growth, relating it to the global crisis in which we are deeply entangled and for which we humans are historically and actively responsible.

References

Balkan-Atlı, N., & Binder, D. (2012). Neolithic obsidian workshop at Kömürcü-Kaletepe (Central Anatolia). In M. Özdoğan, N. Başgelen, & P. Kuniholm (Eds.), *The Neolithic in*

Turkey, New Excavations & New Research, Volume 3: Central Turkey (pp. 71–88). Istanbul: Archaeology & Art Publications

Bar-Yosef Mayer, D. E. (2013). Towards a Typology of Stone Beads in the Neolithic Levant. *Journal of Field Archaeology*, 38(2), 129–42.

Bar-Yosef, O., & Belfer-Cohen, A. (1989). The Levantine "PPNB" Interaction Sphere. In I. Hershkovitz (Ed.), *People and Culture in Change* (pp. 59–72). Oxford: British Archaeological Reports.

Barzilai, O., & Goring-Morris, A. N. (2013). An estimator for bidirectional (naviform) blade productivity in the Near Eastern Pre-Pottery Neolithic B. *Journal of Archaeological Science, 40*(1), 140–147.

Ben-Dor, M., & Barkai, R. (2021). Prey Size Decline as a Unifying Ecological Selecting Agent in Pleistocene Human Evolution. *Quaternary*, 4(1), 7. Retrieved from https://www.mdpi.com/2571-550X/4/1/7

Cessford, C. (2005). Estimating the Neolithic population of Çatalhöyük. In I. Hodder (Ed.), *Inhabiting Çatalhöyük: Reports from the 1995-99 Seasons* (pp. 325–328). Cambridge: BIAA & McDonald Institute for Archaeological Research.

Dunbar, R. I. M. (2022). Managing the Stresses of Group-Living in the Transition to Village Life. *Evolutionary Human Sciences*, 1–39. doi:10.1017/ehs.2022.39

Dunbar, R. I. M., Gamble, C., & Gowlett, J. A. J. (2014). *Lucy to Language: The Benchmark Papers*. Oxford: Oxford University Press.

Dyble, M., Thompson, J., Smith, D., Salali, Gul D., Chaudhary, N., Page, Abigail E., ... Migliano, Andrea B. (2016). Networks of Food Sharing Reveal the Functional Significance of Multilevel Sociality in Two Hunter-Gatherer Groups. *Current Biology*, 26(15), 2017–21. doi:10.1016/j.cub.2016.05.064

Finlayson, B., Kuijt, I., Mithen, S., & Smith, S. (2011). New Evidence from Southern Jordan: Rethinking the Role of Architecture in Changing Societies at the Beginning of the Neolithic Process. *Paléorient*, 37(1), 123–135.

Gamble, C., Gowlett, J., & Dunbar, R. (2014). *Thinking Big: How the Evolution of Social Life Shaped the Human Mind*. London: Thames & Hudson.

Goring-Morris, A. N., & Belfer-Cohen, A. (2011). Neolithization Processes in the Levant: The Outer Envelope. *Current Anthropology*, 52(S4), S195–208. doi:10.1086/658860

Henrich, J. (2015). *The Secret of Our Success: How Culture Is Driving Human Evolution, Domesticating Our Species, and Making Us Smarter*. Princeton, NJ: Princeton University Press.

Hodder, I. (2006). *The Leopard's Tale: Revealing the Mysteries of Turkey's Ancient 'Town'*. London: Thames & Hudson.

Hodder, I. (2011a). Human-Thing Entanglement: Towards an Integrated Archaeological Perspective. *Journal of the Royal Anthropological Institute*, 17(1), 154–77. doi:10.1111/j.1467-9655.2010.01674.x

Hodder, I. (2011b). Wheels of Time: Some Aspects of Entanglement Theory and the Secondary Products Revolution. *Journal of World Prehistory*, 24(2–3), 175–87. doi:10.1007/s10963-011-9050-x

Hodder, I. (2012). *Entangled: An Archaeology of the Relationships Between Humans and Things*. Chichester: Wiley-Blackwell.

Hodder, I. (2014). Çatalhöyük: The Leopard Changes Its Spots. A Summary of Recent Work. *Anatolian Studies*, 64, 1–22. doi:10.1017/S0066154614000027

Hodder, I. (2020). The Paradox of the Long Term: Human Evolution and Entanglement. *Journal of the Royal Anthropological Institute*, 26(2), 389–411.

Hodder, I., & Pels, P. (2010). History Houses: A New Interpretation of Architectural Elaboration at Çatalhöyük. In I. Hodder (Ed.), *Religion in the Emergence of Civilization: Çatalhöyük as a Case Study* (pp. 163–86). Cambridge: Cambridge University Press.

Ibáñez, J. J., Ortega, D., Campos, D., Khalidi, L., Mendez, V., & Teira, L. (2016). Developing a Complex Network Model of Obsidian Exchange in the Neolithic Near East: Linear Regressions, Ethnographic Models and Archaeological Data. *Paléorient*, 42(2), 9–32.

Kuijt, I. (2000). People and Space in Early Agricultural Villages: Exploring Daily Lives, Community Size and Architecture in the Late Pre-Pottery Neolithic. *Journal of Anthropological Archaeology*, 19(1), 75–102. doi:10.1006/jaar.1999.0352

Kuijt, I., Guerrero, E., Molist, M., & Anfruns, J. (2011). The Changing Neolithic Household: Household Autonomy and Social Segmentation, Tell Halula, Syria. *Journal of Anthropological Archaeology*, 30(4), 502–22. doi:10.1016/j.jaa.2011.07.001

Kuijt, I., & Finlayson, B. (2009). Evidence for food storage and predomestication granaries 11,000 years ago in the Jordan Valley. *Proceedings of the National Academy of Sciences of the United States of America,* 106(27), 10966–10970. doi:10.1073/pnas.0812764106

Laland, K. N. (2017). *Darwin's Unfinished Symphony: How Culture Made the Human Mind*. Princeton, NJ: Princeton University Press.

McBrearty, S., & Brooks, A. S. (2000). The Revolution that Wasn't: A New Interpretation of the Origin of Modern Human Behavior. *Journal of Human Evolution*, 39, 453–563.

Mellars, P., & Stringer, C. (1989). *The Human Revolution: Behavioural and Biological Perspectives on the Origins of Modern Humans*. Edinburgh: Edinburgh University Press.

Ortega, D., Ibañez, J., Khalidi, L., Méndez, V., Campos, D., & Teira, L. (2014). Towards a Multi-Agent-Based Modelling of Obsidian Exchange in the Neolithic Near East. *Journal of Archaeological Method and Theory*, 21(2), 461–85. doi:10.1007/s10816-013-9196-1

Özdogan, M., & Özdogan, A. (1999). Archaeological Evidence on the Early Metallurgy at Çayönü Tepesi. In A. Hauptmann, E. Pernicka, T. Rehren, & Ü. Yalçin (Eds.), *The Beginnings of Metallurgy. Proceedings of the International Conference "The Beginnings of Metallurgy", Bochum 1995* (pp. 13–22). Bochum: Deutsches Bergbau Museum.

Pilloud, M. A., & Larsen, C. S. (2011). "Official" and "Practical" Kin: Inferring Social and Community Structure from Dental Phenotype at Neolithic Çatalhöyük, Turkey. *American Journal of Physical Anthropology*, 145(4), 519–30. doi:10.1002/ajpa.21520

Quintero, L., & Wilke, P. J. (1995). Evolution and Significance of Naviform Core-and-Blade Technology. *Paléorient, 21*(1), 17–33.

Rendell, L., Fogarty, L., & Laland, K. N. (2011). Runaway Cultural Niche Construction. *Philosophical Transactions of the Royal Society B: Biological Sciences*, 366(1566), 823–35. doi:10.1098/rstb.2010.0256

Renfrew, C. (1986). Introduction: Peer Polity Interaction and Social Change. In C. Renfrew & J. F. Cherry (Eds.), *Peer Polity Interaction and Social Change* (pp. 1–18). Cambridge: Cambridge University Press.

Renfrew, C. (1996). The Sapient Behaviour Paradox: How to Test for Potential? In P. Mellars & K. Gibson (Eds.), *Modelling the Early Human Mind* (pp. 11–15). Cambridge: McDonald Institute for Archaeological Research.

Renfrew, C. (2008). Neuroscience, Evolution and the Sapient Paradox: The Factuality of Value and of the Sacred. *Philosophical Transactions of the Royal Society B-Biological Sciences*, 363(1499), 2041–7. doi:10.1098/rstb.2008.0010

Renfrew, C., & Dixon, J. E. (1968). Further Analyses of Near Eastern Obsidians. *Proceedings of the Prehistoric Society*, 34, 319–31. doi:10.1017/S0079497X0001392X

Renfrew, C., & Dixon, J. E. (1976). Obsidian in West Asia: A Review. In G. Sieveking (Ed.), *Problems in Economic and Social Archaeology* (pp. 137–50). London: Duckworth.

Roebroeks, W., MacDonald, K., Scherjon, F., Bakels, C., Kindler, L., Nikulina, A., ... Gaudzinski-Windheuser, S. (2021). Landscape Modification by Last Interglacial Neanderthals. *Science Advances*, 7(51), eabj5567. doi:10.1126/sciadv.abj5567

Sterelny, K. (2011). *The Evolved Apprentice: How Evolution Made Humans Unique*. Cambridge: MIT Press.

Sterelny, K. (2018). Religion Re-Explained. *Religion, Brain & Behavior*, 8(4), 406–25. doi: 10.1080/2153599X.2017.1323779

Sterelny, K. (2020). Religion: Costs, Signals, and the Neolithic Transition. *Religion, Brain & Behavior*, 10(3), 303–20. doi:10.1080/2153599X.2019.1678513

Sterelny, K. (2021). *The Pleistocene Social Contract: Culture and Cooperation in Human Evolution*. Oxford: Oxford University Press.

Watkins, T. (2008). Supra-Regional Networks in the Neolithic of Southwest Asia. *Journal of World Prehistory*, 21(2), 139–71. doi:10.1007/s10963-008-9013-z

11 The Problem of Neolithic Religion

Why devote a whole chapter to the subject of religion in the Epipalaeolithic–Neolithic? I would certainly have preferred to treat communal rituals and material symbolism within the context of the evolution of the new kind of cultural niche that Epipalaeolithic–Neolithic represents. But there is a problem: many people, including some archaeologists, assume the existence of particular kinds of religious beliefs and practices in the Neolithic that go well beyond the archaeological evidence. In recent years, the discovery of the monumental buildings populated by dramatically sculpted T-monoliths at Göbekli Tepe have captured worldwide attention and are now commonly proclaimed to be the world's first temples and gods. That is something that has to be challenged: how can such a claim be substantiated? It would require a book in its own right to discuss the different kinds of evidence of rituals and symbolism that might be considered to be religious in intent across the dynamic range of the ENT around southwest Asia. Here I focus on two questions: is there evidence for gods or temples in the Neolithic, and where does the Neolithic fit within theories of the cultural evolution of religion.

At once there is the difficulty arising from the slippery nature of that word religion. Our common ideas about the nature of religion are of our culture and our time; there is no reason to think that such culture-specific contemporary ideas can be applied universally. In short, it is simplistic to read back into prehistoric circumstances 12,000 years ago from our own knowledge and experience of religion. And we cannot simply turn to a dictionary for a definition of the nature of religion. English language dictionaries illustrate the meaning of the word as they have derived it from the way that it is used in (modern and historically recent) literary sources. For example, the Oxford English Dictionary speaks of 'Action or conduct indicating belief in, obedience to, and reverence for a god, gods, or similar superhuman power; the performance of religious rites or observances'. Collins Dictionary is quite straightforward in saying that religion involves 'belief in a god or gods and the activities that are connected with this belief, such as praying or worshipping in a building such as a church or temple'. Chambers Dictionary has searched the literature more widely; its definition avoids mention of a god or gods, and it does not even use the word 'superhuman': 'belief in, recognition of, or an awakened sense of a higher unseen controlling power or powers, with the emotion and morality connected with such'. Dictionaries speak of the present and recent usage of printed

DOI: 10.4324/9781351069281-12

words, but we are concerned with a prehistoric, culturally very different period for which there are no recorded texts.

A second difficulty arises from the view in our increasingly secularised Western world that religion is a distinct category in its own right. That is not how it is seen in many parts of the world, among many (most?) peoples, for whom their faith, whatever it is, is lived through everyday and transcendent life. In a sense, our difficulty in defining 'religion' is self-inflicted. Merlin Donald, in an essay reflecting on Bellah's book *Religion in Human Evolution*, has remarked that, before the Enlightenment, European languages did not need a category corresponding to the modern definition of religion, because there was no so-called 'secular' domain from which it needed to be differentiated; only then did it become important for a new, self-appointed 'secular' élite to distinguish themselves from their predecessors (Merlin Donald 2012: 231).

More than a century ago Emile Durkheim writing about 'The Elementary Forms of Religious Life' defined religion thus: 'Une religion est un système solidaire de croyances et de pratiques relatives à des choses sacrées': 'a religion is a unified system of beliefs and practices relating to sacred matters' (Durkheim 1912, 1915: English edition). Durkheim's perceptive definition has been oft-repeated and generally respected. It is about beliefs and practices, and while it is concerned with 'sacred matters' there is no mention of a god or gods. Durkheim's definition goes on to say that those beliefs and practices hold those who adhere to them together in a moral community. The investigators who have encountered the widest spectrum of religious forms are surely ethnographic fieldworkers; the American anthropologist Clifford Geertz was one of the most widely respected of the twentieth century, and he thought deeply about the religious expressions that he had studied. In an essay on 'Religion as a cultural system' he wrote that a religion is: (1) a system of symbols which acts to (2) establish powerful, pervasive, and long-lasting moods and motivations in men (sic!) by (3) formulating conceptions of a general order of existence and (4) clothing these conceptions with such an aura of factuality that (5) the moods and motivations seem uniquely realistic (Geertz 1966; reprinted in Geertz 1973). More recently, the equally eminent and respected American sociologist Robert Bellah wrote his last, great book on the subject of *Religion in Human Evolution* (Bellah 2011). He paraphrased Geertz's definition as follows: religion is a system of symbols that, when enacted by human beings, establishes powerful, pervasive, and long-lasting moods and motivations that make sense in terms of an idea of a general order of existence. He also discusses at length how religious systems belong to a community whose members share its ideas, forms, and practices, establishing a powerful bond that makes their community. Bellah (2011) and a number of anthropologists who have studied religion (e.g. Baumard & Boyer 2013) emphasise that religion is about practice and experience, rather than a concern for shared beliefs, a point that is emphasised by Kim Sterelny in his important paper on the evolutionary origins of religion (2018: 414, with further references). If religious beliefs and practices serve, as Durkheim proposed, to hold those who adhere to them together in a moral community, then it is the individual's practice that is observable, while beliefs are often difficult to articulate, necessarily personal, and debatable.

We generally think that there is a clear distinction between the natural and the supernatural, but many peoples around the world do not see the world in that way. We make problems for ourselves when we take religion to be an identifiable and separable component within culture. The person who I think helps us to recognise the problem that we cause for ourselves when we make a hard distinction between natural and supernatural is the anthropologist Maurice Bloch. He argues that 'religious-like phenomena' are part of our (modern Homo sapiens) capacity to imagine other worlds (Bloch 2008: 2055–6). Bloch also differentiates what he calls a transactional social world, in which we personally interact with each other, from the uniquely human capacity to attribute transcendental qualities to individuals and groups. The transcendental social means that we have the capacity to see people in terms of their roles such as 'uncle', 'king', 'policeman', or 'ancestor'. Having participated in one of Ian Hodder's inter-disciplinary sessions with the archaeologist-investigators at Çatalhöyük, Bloch responded with a contribution whose title asked 'Is there religion at Çatalhöyük . . . or are there just houses?' (Bloch 2010; Maurice Bloch 2010). Had I been there, I might have replied that a house in the Neolithic could materially demonstrate its transcendental role as 'home' or 'sanctuary' (Watkins 1990). The experience of excavating houses at the site of Boncuklu, which is antecedent to nearby Çatalhöyük, impressed Baird and his colleagues with the repetitive practices, highly structured and symbolically charged domestic activities, ritual and symbolism; they stressed 'the animate and transcendental nature of the house' (Baird et al. 2016).

Our starting point must be taken from the anthropologists and those who have considered the worldwide spectrum of religious practices and beliefs, and not with the dictionary definitions based on recent Western literature. Unfortunately, archaeologists and many others have simply speculated about beliefs in an afterlife to explain what was in the minds of those who buried the dead. Peter Ucko (1969) wrote powerfully about the variety of ethnographic experience and the archaeological interpretation of burials; how could the archaeologist decide which ethnographic analogy was applicable? The same warning should apply when archaeologists and others propose that communities of the Neolithic period built 'temples' where they worshipped 'gods' in the same way that we see around us today, and that we know from recent historical times. But these proposed gods and goddesses of the Neolithic are of great interest – are very exciting – to the general public, which makes them very effective advertising material for the marketing of popular books and of tourism. There is no evidence that would satisfy the great majority of archaeologists, but it is hard to argue effectively against such fantasies.

The thesis of Jacques Cauvin proposed that 'the birth of the gods' at the beginning of the Neolithic occurred ahead of, and was responsible for, 'the beginnings of agriculture' (Jacques Cauvin 1994, 2000). Cauvin's last book has had a much wider readership in the general public than among archaeologists. It might have been better if the two halves of the original French title had been reversed. The book is actually about 'la révolution des symboles au Néolithique' (its French subtitle, which the publisher unfortunately dropped in the English edition), a 'psycho-cultural' revolution in symbolic representation (which we might today describe as a cognitive-cultural co-evolutionary leap): 'the birth of the gods' is a title for the hypothesis that he suggests for the meaning of those symbols. Cauvin had been writing about Neolithic religion and the first gods over many years, but his major book built his arguments for this attractive (for the general public) and controversial (among archaeologists) theory. This is not the place to engage in a critique of the theory. We are concerned only with Cauvin's proposed identification of small female figurines of the earliest Neolithic as representations of a powerful goddess, whose male partner he identified in the figurines of bulls. He identified this male and female pair with gods who are known from the early Mesopotamian world of the third and second millennia BCE. But there is almost no evidence that would support the continuity of those supposed Neolithic divinities across the more than 3000 years that separate the end of the Pre-Pottery Neolithic in the Levant from the early urban civilisation of the Sumerians and Akkadians.

Cauvin's primary objective, I believe, was to show that archaeological facts falsified the theory that the beginning of farming was forced on Neolithic communities when faced with ecological disequilibrium between population and resources (contra Binford and Flannery). Rather, he claimed that the archaeology showed that there was 'a reordering of symbolic material' at the beginning of the Neolithic period, and the beginnings of farming followed later. The dramatic explosion of symbolic representation at the beginning of the Neolithic was evidence, he suggested, of a cognitive change which anticipated the economic change and became manifest within it (I am paraphrasing a passage from a brief essay entitled 'Ideology before Economy' in which Cauvin (2001) sought to explain simply and briefly the purpose of his book.) When it came to explaining the symbolic meaning of the presumed female and male principles, Cauvin says that his proposed model is simply a hypothesis that is open to discussion, improvement in the light better evidence, or modification if the reasoning is found to be weak. We now have much more information than was available to Cauvin thirty and more years ago, not least the exposure of the monumental structures and rich symbolism of Göbekli Tepe, of which Cauvin saw only the beginning. The prominence of human female and bull figurines that so impressed him has been criticised as a false impression. I have to conclude that Cauvin was prescient in identifying a profound and significant change, a revolutionary upsurge, in symbolic material culture that implies new ways of framing their perception of the new world that they were creating; but, at the time when he was writing, he could advance only speculative archaeological evidence. I hope to show that there is a better model. If we want to consider what we might recognise as religion as part of the cultures of the Neolithic in southwest Asia, we need to try a different route, via the kind of cultural evolutionary theory that we considered in Chapter 9.

There is a general observation that there is a spectrum of religious forms just as there is a spectrum from the smallest, mobile foraging societies to our huge and complex contemporary societies. And on the basis of ethnographic and historical observation it has been easy to show that at one end of the spectrum the small and simple mobile forager societies have simple traditions of singing and dancing together, while the emergence of the great world religions can be dated between about 600 BCE and 600 CE in the context of the largest, hierarchically organised social and political systems of those times. While the documentation of the emergence of world religions can be traced and discussed, earlier stages in the evolution of religious ideas, particularly anything earlier than the first literate cultures of early Mesopotamian or Egyptian history, is, to say the least, difficult. The philosopher Karl Jaspers identified the emergence of canonical texts (such as the Hebrew prophets, Amos, Isaiah, and Jeremiah, the central texts of Greek philosophy, Plato and Aristotle in particular, the Analects of Confucius and the Daodejing, Indian texts such as the Bhagavadgita, and the teachings of the Buddha) and the doctrinal world religions as pivotal in human history, calling it the Axial Age (Jaspers 1949; Jaspers & Bullock 1953). The American sociologist of religion, Robert Bellah, has also studied and written extensively of the phenomenon (the last four chapters of Bellah 2011), and several contributors to the book he co-edited with Hans Joas have discussed the immediate precursors of the Axial Age religions (Bellah & Joas 2012). To go beyond the range of surviving written sources we need a different approach.

The question is how religion, within such a definition as those in the earlier paragraphs, may have evolved as an essential element in human cultures. From their work over many years at sites in the Oaxaca Valley in Mexico, Joyce Marcus and Kent Flannery offer us a rare, archaeologically documented sequence, complete with material evidence of increasing religious demands (Marcus & Flannery 2004). Over a period that is quite short in terms of the whole of prehistory, small hunter-gatherer bands were replaced by settled populations who lived in permanent villages and cultivated maize. These in turn gave way to the emergence of a hierarchically organised society whose 'central places' boasted enormous complexes of religious buildings. Ritual paraphernalia were found in individual households and among the public buildings. There were figurines of masked dancers, and examples of pottery masks, as well as costume components such as armadillo shell, crocodile mandible, and macaw wing bones and feathers. So there is plenty of evidence of large-scale collaboration in the construction of ceremonial monuments, and for participation in exciting dancing and painful rituals. In time privileged families emerged who lived in larger houses, practised head-shaping, and gave their dead distinctively rich burials. Their specially produced ceramics were covered in symbols representing the vital forces of Earth and Sky. In the last stage, a new, regional (proto-urban) centre, with an even greater religious centre, replaced the several, competing local centres. And this was the centre where the archaic Zapotec state emerged. Marcus and Flannery note the parallel changes in and increasing complexity of rituals and the scale of special buildings, through small shrines in villages, to full-blown pyramids and the portrayal of extreme ritual practices. The demands of religion grew as the scale and complexity of the community expanded, but it is only in the state-level society that what are recognised as powerful supernatural beings began to be represented.

We cannot assume, of course, that the model of that socio-cultural sequence from central America can serve as a general paradigm for the evolution of religious practices and symbolic expression; but it documents a social and cultural evolutionary pathway that can encourage us in our investigations in southwest Asia. It does illustrate two points that are important for us. On the one hand, it demonstrates what many have observed in general, that the scale and nature of religious practices and beliefs correlates with the scale and complexity of the socio-cultural entity to which it belongs. We are shown how religious behaviours and rituals and their demands can evolve, expand, and grow, changing into new forms that are built from the old ways. And, on the other hand, while an élite stratum that were intimately involved in the religious practices, together with supernatural concepts of the Earth and the Sky as the great forces in the world emerged close to the end of the Formative period, recognisable gods who command service make their first appearance only when Zapotec kingship and the state emerge.

There are two great scholars on whose work I rely. They both offer evolutionary schemes for the emergence and development of religion. Their theses are complementary, and we should start with Robert Bellah's *Religion in Human Evolution: from the Paleolithic to the Axial Age* (Bellah 2011); Kim Sterelny's thesis takes off from Bellah's and we will go there next. Bellah's evolutionary scheme is in three simple stages, each supported by reference to ethnographic analogues. He builds it on the foundations of the cognitive psychologist and neuroscientist Merlin Donald's three-stage scheme that was spelled out in *Origins of the Modern Mind: Three Stages in the Evolution of Culture and Cognition* (1991, 1993 for a precis), and refined and extended in his following book (Donald 2001). Donald proposed three stages in the evolution of human modes of representation that support larger and larger human societies: the first is mimetic (Donald's term), using physical and vocal gesture, followed by the evolution of language (which Donald labels mythic, because of the capacity of language for story-telling and shared narratives), and the third is the externalisation of memory in symbols (external symbolic storage), of which the classic form is in writing (the stage that Donald labels theoretic on account of the fact that the written text offers readers the opportunity for internal analysis and discussion). So Bellah argues that the oldest and the simplest of contemporary human societies have engaged in communal activities that typically involve (mimetic) music and dance. These serve in Durkheim's terms to bind and affirm the solidarity of the group. Bellah brings Donald's modes of symbolic representation with the classification of human societies into the categories of band (typically hunter-gatherer societies), tribe (typically horticulturists, farmers, or pastoralist societies), chiefdom, and state that was first proposed by the anthropologist Elman Service (1962). Thus, in Bellah's scheme the simplest small-scale societies of deep prehistory would have practised singing and dancing rituals, while the early sedentary societies would continue to have their rituals, but would also have traditional stories that told them about their world, their origins and history. The advent of chiefdoms and the early states was associated with the emergence of what Bellah calls 'archaic religion', in other words belief systems that spoke of the relations between people and their gods, in which the rituals were conducted by religious specialists whose knowledge of the beliefs was authoritative, but not the kind of religious systems that characterise the Axial Age. This stage was also dependent on Donald's theoretic stage and the beginning of external symbolic storage of the complex religious beliefs, whether in architecture, monuments, or symbolic imagery. Following Bellah's proposed evolutionary scheme, our sedentary Neolithic communities, with their early stages of cultivation and herding, would fall into his second stage, with rituals and myths. There is no way to accommodate the early Pre-Pottery Neolithic communal buildings, the monumental buildings T-monoliths, sculptures and imagery of Göbekli Tepe in that second stage.

My second authority is Kim Sterelny, who has recently devoted much thought to the subject of the evolution of religion (Sterelny 2018, 2020, 2021). He has the advantage over Bellah in that he has been working at the forefront of cultural evolutionary theory, he knows a good deal of prehistory (where Bellah relied mostly on ethnographic information), and he has been able to take into account a considerable body of work published since Bellah's big book. I am following Sterelny's analysis of the problem of developing an account of the evolution of religion within the framework of the evolution of human social life, and I will differ from his thesis only when it reaches the Neolithic period, where I will dare to qualify his conclusions.

Sterelny illustrates the problem that is posed by religion for any evolutionary account with a quotation from an essay by Maurice Bloch:

> How could a sensible animal like modern Homo sapiens, equipped by natural selection with efficient core knowledge . . . , i.e., knowledge well suited for dealing with the world as it is, hold such ridiculous ideas as: there are ghosts that go through walls; there exist omniscients; and there are deceased people active after death?
>
> (Bloch 2008: 2055)

Sterelny notes that music (singing, dancing, and playing music) and story-telling are, like religion, practically universal in all human societies, and they do not, on the whole, seem to be activities that are essential to staying alive or reproduction. His thesis is that religion has piggy-backed onto the much more ancient practices of music (singing and dancing), and story-telling (or mythology); practice and experience were the operative characteristics, before belief entered the equation. In short, 'early, *embodied* religion transitioned into *articulated* religion, which in turn, in some cultures and contexts, transitioned into *ideological* religion' (Sterelny 2017: 14). He describes his thesis as a tweak on those of Hayden (2003), Whitehouse (2016), and especially Bellah (2011); in following Sterelny, whose thesis is supported by many of those who know best, I am in good company.

His argument is that ancient, embodied religion (echoing Donald's mimetic stage) was nothing but a distinctive, locally specific package of rituals, ceremony, collective activity, a mechanism of internal affiliation and identity. As mobile forager societies became larger and more complex (late in the Pleistocene, between 100 and 50 thousand years ago), embodied became articulated religion (again, echoing Donald's mythic stage) through the addition of reliably transmitted, mythopoetic narrative that was typically concerned with who we are, why we are here, and our right to our place in the world. When articulated religion becomes ideological (in Donald's terms, theoretical), the narrative is much more elaborated, it is compulsorily believed as true, is moralised, and ideology becomes central in its role. In his most recent exploration of the subject (Sterelny 2021: Chapter 4), he argues that ideological religion, characterised by costly rituals, emerges in the Neolithic period, when it is associated with the establishment of complex hierarchical social worlds. His paradigmatic case is Göbekli Tepe, with its great circular buildings and vivid symbolic imagery, which he describes as 'an extreme manifestation of competition for prestige within a segmented society with access to social surplus' (Sterelny 2021: 147). I cannot follow him in detail in his diagnosis of early Neolithic social worlds as 'complex' and 'hierarchical', 'with access to social surplus'. Those terms 'complex' and 'hierarchical' are variously used, which makes them hard to define; I am sure that Sterelny would be able to explain how he understands them, but they are commonly used to describe the early urban societies of protohistoric Mesopotamia or early Egypt, with their strong distinctions of wealth and power.

Both Bellah ('archaic religion') and Sterelny ('ideological religion') refer back to Merlin Donald's 'theoretic' stage, which Bellah associates with the emergence of hierarchically organised societies, such as chiefdoms and early states, very different from the societies that we have seen in the Neolithic of southwest Asia. Sterelny, however, directly associates the emergence of 'ideological religion' with the emergence of the Neolithic in southwest Asia and the existence of 'a segmented society with access to social surplus' that is engaged in competition for prestige. In the previous chapter, I sought to define the Epipalaeolithic–Neolithic period as pivotal within the long-term of human cultural evolution. These were no longer small-scale egalitarian societies, but neither were they large-scale societies with institutionalised social and political power of the kind that emerge in the Uruk period in Mesopotamia. The Neolithic societies were transegalitarian, and there were surely individuals with prestige, who were relied upon for their particular expertise, skill, or wisdom; but there is no sign of institutionalised status. Kim Sterelny refers specifically to the monumental structures and imagery of Göbekli Tepe as evidence for competition for prestige, implying that leading figures within a segmented society could enhance their prestige by means of coordinating the creation of such massive buildings, populated with anthropomorphic monoliths of superhuman scale. My response is that the communal buildings of the early Pre-Pottery Neolithic, including the Göbekli Tepe buildings, required the expert knowledge and skills of a team of several people: in the example of the Göbekli Tepe buildings and sculptures, their careful design and layout required surveying and measuring skills (Haklay & Gopher 2020), the symbolic significance of their design and furnishing was another matter, and there had to be skilled team leaders to manage the quarrying, carving, transport and erection of the massive monoliths. I would rather suppose that the great communal buildings were testimony to the effective teamwork of the community, led perhaps by someone who was the recognised authority on the ideology of their shared beliefs.

The design, construction and functioning of the massive circular stone buildings of Göbekli Tepe, with their formal settings of carved anthropomorphic monoliths, implies leadership and management, and not just in design and execution of the structures, and the logistical organisation of the work. It is clear that the great buildings at Göbekli Tepe continued to be reshaped, remodelled, and used over centuries, which implies that those with the leadership and organisation skills, as well as the ritual and mythological knowledge, continued to be of central importance to the community. We have no direct knowledge of those people, but they were, I suggest, different from others whose knowledge and skills earned them prestige and respect in a particular field, for example, as a mid-wife, or as the maker of fine chipped stone tools. If the practices and beliefs were the expressions of the cultural memory and solidarity of their society, as I argued earlier in this chapter, then those whose knowledge and skills in the ritual field merited the prestige and the general respect of their communities. But there is no sign in the archaeology of their existence as an élite who lived in different houses, had special possessions, or were buried in special places or special ways.

In terms of the cultural evolutionary timescale, things began to move faster with the Epipalaeolithic, and symbolic activity and ritual practice came into the foreground in the final Epipalaeolithic and much more markedly in the early Pre-Pottery Neolithic. And as society and the economy changed in the Neolithic, so did rituals and acts of symbolic expression. But nowhere are there signs of the emergence of ideas of supernaturally powerful gods, or of an institutionalised élite acting as priestly intermediaries on behalf of their communities. For changes of that kind it is necessary to move on to the fourth millennium BCE, as the recent essays of Piotr Steinkeller (2017) and the lengthy essay on the Uruk phenomenon by Gerd Selz (2020) illustrate.

In short, the first signs of religions that involved the service of powerful gods by an institutionalised temple religious-political élite, the advent of what Robert Bellah called 'archaic religion' are to be found millennia beyond the Neolithic period with which we have been dealing. The evidence for such religious beliefs, practices, and symbolic representation is lacking in the Neolithic, but, since the period may be described as transegalitarian in several ways, perhaps we can suppose that, as in the Meso-American example described by Marcus and Flannery, there were already ideas, images, and narratives that spoke of a supernatural order and forces in the world, though they were not yet institutionalised.

In keeping with the idea that Neolithic societies in southwest Asia were transegalitarian – engaged in the dynamic transition from egalitarian into socio-economically hierarchical societies – I suggest that we can think of their religious ideas and practices as transitional: in Donald's terms, that is 'mythic' but engaged with an emerging 'theoretic' mode. The ideology is not embedded in sacred writings curated by a priesthood, but is signified in architecture and imagery (Watkins 2015, 2017). In Sterelny's terms, the architecture, sculpture and imagery embodied and materially articulated an ideology. The human figures in the T-monoliths of Göbekli Tepe cannot be identified as 'gods'; they are imagined as anthropomorphic figures, but there is no reason to think that they are gods. And the monumental circular structures do not seem to be designed for the ritual service of the gods, which would typically require a place laid out for the deity to await the service of the priest, servant or worshipper. The Göbekli Tepe buildings have become iconic, but they are architecturally like the other 'communal buildings' that we met in the early Pre-Pottery Neolithic. The common features seem to be their enclosed nature, and, in a number of examples, a 'bench' at the base of the wall, as if inviting a relatively small number of people from the community to meet for whatever purpose. They are unlike the other communal buildings in that the anthropomorphic T-monoliths portray an assembly of figures attending upon two figures who stand side by side in the centre. The assembly of figures may have been, for example, imaginative representations of the archetypal ancestors, the founders of lineages, or the assembly of community elders of long tradition; I have suggested that they were in some way 'theatres of memory' that accommodated ritual assembly and celebrated the community's history (Watkins 2004). We do not have contextual information that could help us to define the transcendental identity of the schematic figures engaged in the drama of that assembly, and therefore it is not possible to say they are gods.

Thus, while there may have been ideas of supernatural beings (transcendent as opposed to transactional, as Bloch would define them), there is no reason to think that there are representations of 'gods', that is superhuman agents who could act in the world, in the Neolithic (Watkins 2019). Earlier I said that I followed the thinking of Kim Sterelny on the evolution of religion; his ideas were built on the work of Bellah, whose major book I greatly admired, and whose evolutionary framework for religion was taken from the theory of Merlin Donald, which I have employed in relation to the making of meaning in Neolithic architecture (Watkins 2015, 2017). I am happy to follow Sterelny's lead: he has concluded that 'the increase in scale and complexity [in the Neolithic of southwest Asia] triggered the beginnings of a transition to ideological religion' (Sterelny 2020: 15). Yes, it may well be the beginnings of that transition; but ideological religion, in Sterelny's terms, 'with its formalised and compulsory doctrine, specialised priesthood, and sharp distinction between the laity and religious élite', can only be seen to emerge later at the borderline between prehistory and the beginnings of history.

It is very unsatisfactory for me to have to finish this chapter (a) with a negative conclusion – that there is no reason to identify either gods or temples in the early Neolithic – and (b) to be unable to offer an alternative interpretation of the iconic monumental architecture and vivid sculptured imagery of Göbekli Tepe. But archaeology is always in difficulty with the unique instance, as with the Göbekli Tepe monumental buildings. We can look forward to having a better context for these discussions as Turkish archaeologists pursue the excavation of Karahantepe (Karul 2021) and other sites in the area where there are known to be T-monoliths. And it is to be hoped that further field research will produce information about the Epipalaeolithic that precedes Göbekli Tepe, and the later Pre-Pottery Neolithic that succeeds the window of our limited perspective. Till now attention has been focused on that cluster of large buildings and their T-monoliths, and we have almost no contemporary cultural context within which to understand them. We also need to know something of how the Göbekli Tepe phenomenon developed out of the preceding period, as well as something of the developments that followed from the closure of the monuments. Finally, on the difficult matters of religion and the evolution of religion, and equally on the large gaps in our archaeological understanding of sites like Göbekli Tepe, we must hope that continuing discussions will help us to progress our ideas, eliminate what is shown to be erroneous, and carry forward our knowledge of articulated and approaching ideological religion in the Neolithic.

References

Baird, D., Fairbairn, A., & Martin, L. (2016). The Animate House, the Institutionalization of the Household in Neolithic Central Anatolia. *World Archaeology*, 49(5), 753–76. doi:10.1080/00438243.2016.1215259

Baumard, N., & Boyer, P. (2013). Explaining Moral Religions. *Trends in Cognitive Sciences*, 17(6), 272–80. doi:10.1016/j.tics.2013.04.003

Bellah, R. N. (2011). *Religion in Human Evolution: From the Paleolithic to the Axial Age.* Cambridge, MA: Belknap Press of Harvard University Press.

Bellah, R. N., & Joas, H. (2012). *The Axial Age and Its Consequences.* Cambridge, MA: Belknap Press of Harvard University Press.

Bloch, M. (2008). Why Religion Is Nothing Special but Is Central. *Philosophical Transactions of the Royal Society B-Biological Sciences*, 363(1499), 2055–61. doi:10.1098/rstb.2008.0007

Bloch, M. (2010). Is There Religion at Çatalhöyük . . . or Are There Just Houses? In I. Hodder (Ed.), *Religion in the Emergence of Civilization: Çatalhöyük as a Case Study* (pp. 146–62). Cambridge: Cambridge University Press.

Cauvin, J. (1994). *Naissance des divinités, naissance de l'agriculture: la révolution des symboles au Néolithique.* Paris: CNRS editions.

Cauvin, J. (2000). *The Birth of the Gods and the Origins of Agriculture.* Cambridge: Cambridge University Press.

Cauvin, J. (2001). Ideology before Economy. *Cambridge Archaeological Journal*, 11(1), 106–7.

Donald, M. (1991). *Origins of the Modern Mind: Three Stages in the Evolution of Culture and Cognition.* Cambridge, MA; London: Harvard University Press.

Donald, M. (1993). Précis of Origins of the Modern Mind: Three Stages in the Evolution of Culture and Cognition. *Behavioral and Brain Sciences*, 16(4), 737–48. doi:10.1017/S0140525X00032647

Donald, M. (2001). *A Mind So Rare: The Evolution of Human Consciousness.* New York, NY: Norton.

Donald, M. (2012). The Complex Origins of Religion: The Work of Robert Bellah. *Religion, Brain & Behavior*, 2(3), 230–7. doi:10.1080/2153599X.2012.721216

Durkheim, E. (1912). *Les formes élémentaires de la vie religieuse.* Paris: Librairie Félix Alcan.

Durkheim, E. (1915). *The Elementary Forms of the Religious Life: A Study in Religious Sociology (Translated by J. W. Swain).* London: Allen & Unwin.

Geertz, C. (1966). Religion as a Cultural System. In M. Banton (Ed.), *Anthropological Approaches to the Study of Religion* (pp. 1–46). London: Tavistock Publications.

Geertz, C. (1973). *The Interpretation of Cultures: Selected Essays*. New York, NY: Basic Books.

Haklay, G., & Gopher, A. (2020). Geometry and Architectural Planning at Göbekli Tepe, Turkey. *Cambridge Archaeological Journal*, 1–15. doi:10.1017/S0959774319000660

Hayden, B. (2003). *Shamans, Sorcerers, and Saints: A Prehistory of Religion*. Washington, DC: Smithsonian Books.

Jaspers, K. (1949). *Vom Ursprung und Ziel der Geschichte*. Zürich: Artemis-Verlag.

Jaspers, K., & Bullock, M. (1953). *The Origin and Goal of History*. London: Routledge & Kegan Paul.

Karul, N. (2021). Buried Buildings at Pre Pottery Neolithic Karahantepe/Karahantepe Çanak-Çömleksiz Neolitik Dönem Gömü Yapıları 2021. *Türk Arkeoloji ve Etnografya Dergisi*, 86, 19–31.

Marcus, J., & Flannery, K. V. (2004). The Coevolution of Ritual and Society: New 14C Dates from Ancient Mexico. *Proceedings of the National Academy of Sciences*, 101(52), 18257–61. doi:10.1073/pnas.0408551102

Selz, G. J. (2020). The Uruk Phenomenon. In K. Radner, N. Moeller, & D. T. Potts (Eds.), *The Oxford History of the Ancient Near East – Volume I: From the Beginnings to Old Kingdom Egypt and the Dynasty of Akkad*. Oxford: Oxford University Press.

Service, E. R. (1962). *Primitive Social Organization: An Evolutionary Perspective*. New York, NY: Random House.

Steinkeller, P. (2017). *History, Texts and Art in Early Babylonia: Three Essays*. Berlin: De Gruyter.

Sterelny, K. (2017). Artifacts, Symbols, Thoughts. *Biological Theory*, 12(4), 236–47. doi:10.1007/s13752-017-0277-3

Sterelny, K. (2018). Religion Re-Explained. *Religion, Brain & Behavior*, 8(4), 406–25. doi:10.1080/2153599X.2017.1323779

Sterelny, K. (2020). Religion: Costs, Signals, and the Neolithic Transition. *Religion, Brain & Behavior*, 10(3), 303–20. doi:10.1080/2153599X.2019.1678513

Sterelny, K. (2021). *The Pleistocene Social Contract: Culture and Cooperation in Human Evolution*. Oxford: Oxford University Press.

Ucko, P. J. (1969). Ethnography and Archaeological Interpretation of Funerary Remains. *World Archaeology*, 1(2), 262–80. Retrieved from www.jstor.org/stable/123966

Watkins, T. (1990). The origins of house and home? *World Archaeology* 21(3), 336–347.

Watkins, T. (2004). Architecture and 'Theatres of Memory' in the Neolithic of South West Asia. In E. DeMarrais, C. Gosden, & C. Renfrew (Eds.), *Rethinking Materiality: The Engagement of Mind with the Material World* (pp. 97–106). Cambridge: McDonald Institute of Archaeological Research.

Watkins, T. (2015). Ritual Performance and Religion in Early Neolithic Societies. In N. Laneri (Ed.), *Defining the Sacred: Approaches to the Archaeology of Religion in the Near East* (pp. 153–60). Oxford: Oxbow Books.

Watkins, T. (2017). Architecture and Imagery in the Early Neolithic of Southwest Asia: Framing Rituals, Stabilizing Meanings. In C. Renfrew, I. Morley, & M. Boyd (Eds.), *Ritual, Play and Belief in Early Human Societies* (pp. 129–42). Cambridge: Cambridge University Press.

Watkins, T. (2019). When Do Human Representations Become Superhuman Agents. In J. Becker, C. Beuger, & B. Müller-Neuhof (Eds.), *Iconography and Symbolic Meaning of the Human in Near Eastern Prehistory* (pp. 225–35). Vienna: OREA, Austrian Academy of Sciences Press.

Whitehouse, H. (2016). Cognitive Evolution and Religion: Cognition and Religious Evolution. *Issues in Ethnology and Anthropology*, 3(3), 33–47.

12 The Triple A

Aggregation, Acceleration, Anthropocene

We have been examining a transformative process across a prehistoric period of around fifteen millennia at the end of the Pleistocene and the beginning of the Holocene periods: how a new kind of cultural niche began to speed up cultural evolution, intensifying the way that sedentary communities became larger, how they came to rely on mixed farming, and how they intensified their social networks of interaction. At the end of our story, we saw how there was another, unexpected transformation, the beginnings of the rapid geographical expansion of farming communities in all directions beyond the core area in southwest Asia. As a long-term member of an international community of researchers, I have been trying to play a part in learning about an important chapter in deep human history. That has been the subject of this book, but in this final chapter I want to take a different perspective.

It is easy to think that some historical perspective is essential to understanding who we are and how our world has come to be as it is today. Historians seeking to explain a contemporary situation may refer back to the European Enlightenment, the industrial revolution, the American or French revolutions, the global conflicts of the first half of the twentieth century, the world post-1950, or to the date of the election of the present government. Historical perspectives on our present condition have become more and more confined; it is common to find books that look no further back than the middle of the last century, and recently I have been reading books that relate the global climate crisis to the accelerating rates of change over the last 20 or 30 years. We are assailed with increasing urgency by the warnings of the world's leading environmental scientists, climatologists, oceanographers and biologists of the alarmingly rapid changes in climate, global warming, sea-level rise, loss of biodiversity, and their potentially disastrous effects on our world. But there is good reason to be interested in a time that is more than 10,000 years in the past, and a region which, for most of us, is unfamiliar.

That is the message of the Anthropocene. Struck by the realisation of the huge and rapidly growing impacts of human activities on the Earth and its atmosphere, it seemed more than appropriate to the Nobel Prize winning atmospheric chemist Paul Crutzen and the biologist Eugene Stoermer to emphasise the central role of human societies on Earth's atmosphere, oceans and ecology by proposing the term Anthropocene for the current geological epoch, for whose characteristics human

DOI: 10.4324/9781351069281-13

activities must take responsibility (Crutzen & Stoermer 2000). They acknowledged the difficulty of identifying the onset of the 'Anthropocene', but proposed the latter part of the eighteenth century, when the global effects of human activities became clearly identifiable in the analyses of air trapped in polar ice. That showed rapidly increasing global concentrations of carbon dioxide and methane, coinciding with James Watt's design of an efficient steam engine in 1784 and the rapid industrial implementation of coal-fired power (Crutzen 2002: 23). The upturn in the graph of concentrations of CO_2 in the atmosphere was clear. The picture has become more complicated as other disturbing changes have been tracked. Year by year the warnings have become more urgent that we are reaching, and passing, the tipping points that may lead to irreversible climate system collapse (e.g. Lenton et al. 2008; Rockström & Gaffney 2021; Steffen et al. 2018).

There has been much discussion of the Anthropocene since the term was first proposed; indeed, a scientific journal of that name was soon established. The subject attracts the interest of many scientists beyond geology, because of course we are all becoming aware of the urgency and the complexity of global warming. Rapid advances in scientific techniques and increasing research focus have accelerated our knowledge in the last two or three decades. The acceleration of the science has only served to emphasise the accelerating pace of global warming. In those terms, fixing the start of the Anthropocene at the dawn of steam-powered industry and transport may seem an irrelevant academic argument. But the Anthropocene is not something that was inaugurated by the industrial revolution; the anthropogenic release of carbon dioxide into the atmosphere did not begin in the 1780s, and atmospheric concentrations of carbon dioxide is only one of a number of anthropogenic components of the Anthropocene. (A useful summary of the debate can be found in the introduction to a special issue of *Anthropocene*: Tarolli et al. 2014.)

Anthropogenic landscapes now cover most of Earth's land surface, not to mention the environmental and ecological impacts of industrialisation and urbanisation. At one level, the Anthropocene is the effect of human-landscape interaction since the beginnings of cultivation, herding, and wholesale landscape and ecosystem manipulation. As human populations dependent on farming have grown exponentially over the millennia, so the growth of demand for agricultural land and the intensity of agricultural production have expanded in parallel. Two leading archaeo-zoologists have proposed that the onset of farming is in effect the inception of the Anthropocene (Smith & Zeder 2013); they emphasise how the beginnings of farming represent a transformation of the human niche in its environmental, social and cultural contexts, enabling human societies to modify species and ecosystems and manage extensive areas around their settlements for the production of the food for the densest human populations in history, as well as for the provision of other essential materials such as wood for building, cooking and heating. By the end of the Neolithic period much of southwest Asia was peppered with settlements surrounded by landscapes of fields and areas of intensively grazed land. How changed it was from how the region looked when the last Upper Palaeolithic hunter-gatherers became the first hunter-harvesters of the Epipalaeolithic period 15,000 years earlier. And within a millennium or so of the end of the Neolithic in southwest Asia much of Europe's landscape was also peopled with farming populations with similar consequences for the natural environment.

The way of life that developed out of the ENT in southwest Asia, together with the equivalent independent transformations that quickly followed in short order in other parts of the world, set in process the acceleration that characterises the Anthropocene in terms of expanding human population, the extent of land that was taken for farming with the same direct consequences in terms of reduced biodiversity and deforestation. The process of those changes from 12,000 years ago until the present have been assessed and charted (Stephens et al. 2019). Their comprehensive picture of the expansion of human land use showed that humans had transformed the face of Earth by 1000 BCE, significantly earlier and to a greater extent than had been appreciated by Earth scientists. The environmental scientist William Ruddiman (2017, 2010, 2013) has shown that, however small the atmospheric impact of the beginnings of cultivation and herding that we have been considering, the intensification and extensification of farming is traceable in amounts of carbon dioxide and methane both in Greenland and Antarctic ice cores from at least five or six thousand years ago. There is an appreciable acceleration in the accumulation of methane in the atmosphere from that time, which Ruddiman attributes to carbon dioxide-emitting early forest clearance in Europe and China, the increased keeping and use of methane-emitting cattle, and methane-emitting rice irrigation across south and east Asia. And those inter-related processes, which steadily became more complex and, because of the increase in human population, more intensive, has now come to a global crisis.

Beyond the Impact of Farming and Land Use

Another factor in the growth of the Anthropocene is the exponential growth of the global human population. We are very aware of the scale of the global population, and we are often told how rapidly that population has grown over the last century, or since the industrial revolution. But any exponential growth starts very slowly, almost imperceptibly; and there are clear signs, as we saw in Chapters 7 and 8, that Neolithic populations expanded at rates that were without precedent in prehistory. The significant factor behind that Neolithic demographic transition (Bocquet-Appel & Bar-Yosef 2008) was the new form of the Neolithic cultural niche, which encouraged large numbers of people to live together in permanent settlements that were nodes in extensive and intensive socio-cultural networks.

We have met the terms Anthropocene and acceleration: now we should turn our attention to aggregation. As we saw in Chapter 4, and as again mentioned in Chapter 10, Epipalaeolithic specialists have been exploring what they have called aggregation sites in wetland areas in the semi-arid interior of Jordan (Garrard & Byrd 1992; Jones et al. 2016; L. Maher 2017); and some Neolithic specialists have also mentioned the word 'aggregation' in connection with the coming together of people into permanent settlements. Hence, I was intrigued when I came across 'aggregation' in the context of papers in major science journals on the subject of 'urban science' and 'settlement scaling theory'. Of course, since at least the industrial revolution there have been social scientists, economists, geographers, architects and philosophers investigating and discussing the extraordinary phenomenon of cities and urbanism. Urban studies have come a long way since Gordon Childe introduced the idea of an urban revolution and described how the first urban centres in southern Mesopotamia concentrated large populations among whom full-time specialists served each other's needs in a complex economy that was administered by literate scribes and ruled by kings who could command armies. The intensified level of urban and rural population interacted with the intensified and extensified levels of economic activity to push upward the graph of accelerating impact on the world's resources and the effects on the atmosphere. Urban studies have come a long way since Childe's time, of course, studying contemporary urbanism as engines of creativity and innovation, wealth creation and economic growth, as well as sources of pollution and disease. It has been difficult to accumulate good data on any of the various fields, and the increasingly rapid growth and spread of urbanism has emphasised the urgent need for 'an integrated, quantitative, predictive, science-based understanding of the dynamics, growth and organisation of cities' (Bettencourt & West 2010: 912). In the first decades of the present century so much data has been collected and standardised and is openly accessible; information technology has recently opened the way to 'big data' that could be assembled from everywhere and (for those with the mathematical and computer science skills) analysed at your desk.

A decade ago Luis Bettencourt (2013) announced a new approach to the analysis of urban societies. All sorts of factors that are essential elements of life in cities are shown to scale in relation to population, but they do not necessarily have a simple linear relationship: they may be scaled according to super-linear (or sublinear) relations. For example, a settlement with a population of 10,000 is 10 times larger than a population of 1000; but settlement scaling analysis might find that the population of 10,000 is not 10 times, but 12 times more productive than the smaller settlement, while it occupies space and requires resources at only about eight times the population; productivity therefore is scaled at an exponent of 1.2 relative to population, while the exponent for economies of scale is 0.85 relative to population. It has been observed by sociologists and anthropologists that these advantages are general and are independent of formal markets, modern administrative or political structures, wages and salaries, or levels of industrialised production. The more a population can aggregate, the greater their potential productivity: could this aggregation possibly apply to the pre-urban Neolithic?

In a typical table of results relating to the analysis of contemporary US city-regions (Bettencourt et al. 2007: Table 1), there are factors that simply scale in a linear fashion – more population requires more housing for example; there are factors where the economy of scale means that the scaling exponent is sub-linear; and there are factors that scale in a super-linear fashion, signalling innovations, wealth and productivity, but also, at least in modern cities, criminality and poverty. The analyses undertaken by the 'Social Reactors' group working on settlement scaling theory have shown that a very similar scaling component applies both to contemporary US city-regions, and to cities all around the world, and to early and ancient cities, whether in central America or early Mesopotamia (Lobo, et al. 2020; Ortman & Coffey 2017; Ortman et al. 2016; Smith 2019).There are costs as well as benefits to urban life, but the benefits of aggregation and settlement scaling outweigh the costs. Michael Smith, one of the collaborating group, has borrowed the phrase 'energised crowding' to label the intensifying effects of aggregation and living together in large numbers (Smith 2019). In common with a number of researchers from various disciplinary backgrounds Smith talks about scalar stress; living close together in large numbers is stressful, and the stresses increase with the scale of the lived environment. Consequently, societies need norms of behaviour and structuring institutions for the community in order to counter scalar stress. But Smith argues that the 'energised crowding' of urban situations has a role in generating economic growth, and generally the benefits outweigh the costs.

Two members of the group, Ortman and Coffey, set out to test whether the same scaling phenomenon applied to pre-urban societies, using data from pre-Columbian pueblo settlements of the Central Mesa Verde in Mexico and settlements of the northern Middle Missouri regions of North America, dating between the sixteenth and eighteenth centuries CE (Ortman & Coffey 2017). Their results indicate that economies of scale with respect to settlement area and increasing returns to scale with respect to house area were found in both cases. And they conclude that the properties observed derive from the same social networking processes that lead to scaling phenomena in modern or earlier cities. The two case-studies are of 'middle-range societies' that are certainly smaller and simpler than most urban societies, but they are still well-removed from our earliest Neolithic settlements. Two questions arise therefore. Does the settlement scaling phenomenon apply to all sedentary populations in permanent settlements, or do the effects of aggregation begin to appear only at a certain stage in the growth of settled communities? Can the analyses developed by settlement scaling theory be applied to our data from the several millennia of Neolithic settlements in southwest Asia with the same effects as have been found by the Social Reactors Project?

Across the ENT we see the implementation of aggregation replacing the mobile forager strategy, and we can certainly see the pace of demographic and cultural change accelerate. As mentioned earlier in this chapter, in wetland areas in semi-arid northeast Jordan there are 'aggregation sites' in the early and middle Epipalaeolithic. Lisa Maher (2017) argues that the extensive and stratified site of Kharaneh IV was repeatedly occupied for substantial periods (months? seasons?) by large numbers of people coming together from a wide region to the west. Maher and Macdonald (2013) discuss the massive amounts of chipped stone material, the variability of which, they suggest, represents a mixture of knapping traditions from distinct groups. We are seeing the traces of many different groups congregating, living together seasonally, interacting, sharing and exchanging material goods and knowledge, and generally building their social relations with one another (Maher 2017: 685).

From the late Epipalaeolithic and early Pre-Pottery Neolithic, settlements became established that were built, rebuilt and occupied by sedentary communities of people who lived together in large numbers (relative to their Palaeolithic ancestors) generation after generation often over many centuries. The new way of life implies new norms and institutions, the means to counter the different causes of scalar stress (see Dunbar 2022 for a probing discussion of those social stresses). The norms and institutions that framed the new social and cultural way of life were articulated, as we have seen in earlier chapters, in the material life of the community, whether in architecture, monumental communal buildings, imagery or rituals. The material life of the community took place in and around the buildings that they had made, which framed and gave meaning to their social lives. Ian Kuijt (2000) charted for the southern Levant the increasing population densities in settlements across the range of the Pre-Pottery Neolithic, and there is an increasing range of size of settlements; over the three and a half millennia it seems that the density of houses grows exponentially, as does their overall size and internal compartmentalisation; settlements become larger, some of them much larger, leading to the late Pre-Pottery Neolithic 'mega-site' phenomenon. There is also evidence for cumulative culture in the increasing range of craft products, skills and activities, together with increasing range and intensity of exploitation of resources, and the exchange of materials and products.

The kind of data necessary for settlement scaling analysis is certainly available, at least for some sites. A foretaste can be found in a chapter whose purpose was to contextualise a change in the mode of religiosity at Çatalhöyük and its role in the evolution of social complexity (Whitehouse et al. 2014). There is a simple graph (Figure 6.3) that shows the overall growth in the internal area of houses through two millennia at Çatalhöyük, and we know that there was also increasing variation both in the scale and in the elaboration of the furnishing of houses. The costs of aggregation may have increased as the population grew, living in closely packed houses with almost no public spaces, but so did cultural cumulation. The authors note the beginnings of ceramic technology, which was a diversification of the already present fired clay technology. In turn the domestic use of ceramics implies developments in food processing, and also of its presentation. The authors also note the increasing specialisation in the production of ground stone tools in the latest levels, increased specialisation in obsidian tool-making, and increase in the number and range of clay balls, clay tokens, and stamp seals in the late levels.

At Çatalhöyük, Aşıklı Höyük (Yalçin & Pernicka 1999) and other sites, especially Çayönü in southeast Turkey (Maddin, et al. 1999; Özdogan & Özdogan 1999; Roberts et al. 2009), there have been finds of small objects made of native copper. Çayönü was close to a major copper resource at Ergani, and examination of the multiple small awls, hooks, and rings has shown that they were worked hot, and not simply cold-hammered. At Tell Halula, a later Pre-Pottery Neolithic settlement on the Euphrates in north Syria, 11 of the more than 100 burials that were found included copper or lead beads (Molist et al. 2009). There was also a crescentic ornament of native copper that had been annealed between spells of cold-hammering. Sites like Tell Halula and Aşıklı Höyük do not have convenient sources of copper in their vicinity, so we can infer that copper or copper objects were being exchanged in the later Pre-Pottery Neolithic period, and in some places at least there was technological knowledge of how to heat-treat the metal to prevent it becoming brittle under cold-hammering. The making of lime plaster, which was very exten-sively used on walls and floors throughout the Levant in the later Pre-Pottery Neo-lithic involved another application of pyrotechnology, processing by fire (Goren & Goring-Morris 2008). The burials at Tell Halula have also produced a consider-able number of beads and personal ornaments in a variety of materials. Danielle Bar-Yosef Mayer (2013) has noted that, whereas marine shells were extensively used for beads in the Upper Palaeolithic period, at the end of the Epipalaeolithic the first stone beads appear. Through the Neolithic, the diversity of shapes, raw materials, and production methods of stone beads increases. She concludes that, by the later Pre-Pottery Neolithic, there were likely workshops that specialised in the production of particular types of beads, which were then disseminated through the networks of social exchange.

Last but by no means least, we should not overlook the new skills and multiple areas of knowledge that were involved in the cultivation of crops and the mainte-nance of herds of animals. There is an enormous range of knowledge and skills in-volved in even the simplest mixed farming. It is often glibly said that the Neolithic was the foraging to farming transition, when people became farmers and stopped being hunter-gatherers. But, as we have seen, that is a dangerously simplistic slo-gan: while Neolithic communities gradually increased their reliance on cultivated crops and herded animals, they continued to hunt and trap, and to forage for plant foods. The skills and knowledge required for cultivation and herding were exten-sions of and additions to their hunting and gathering skills and their knowledge of plants and animals.

While aggregation may enhance cultural and economic productivity, no settle-ment, whether a twenty-first century urban metropolis or a Pre-Pottery Neolithic community, functions in isolation and networking is also important. In another paper members of the Social Reactors Project looked at a group of settlements in the Peruvian Andes and how things changed as they were brought into the dynamic social and economic network of the Inka Empire around 1450 CE (Ortman et al. 2016). The study of the material outputs across the settlements and across house-holds shows that the region experienced a marked economic growth as a result of the intensification of human social connectivity and material flows within the Inka imperial network.

Settlement scaling theory functions within networks. And the importance of networking has been illustrated for us in the important work done by Juan José Ibáñez, David Ortega and colleagues on the crescendo of distribution of central Anatolian obsidian through the Pre-Pottery Neolithic of the Levant (Ibáñez et al. 2015; Ortega et al. 2014). Their statistics document the intensification of social exchange through the Pre-Pottery Neolithic by means of the growth in the amounts of obsidian in the network. And their simulation studies showed that the working of the exchange networks became more complex and sophisticated with time. People did not necessarily rely on exchanges with people from the nearest neighbour com-munity; rather, they tended to concentrate on building relations with larger settle-ments up to 180 km away. There is a particularly interesting feature of their study in the table of the percentages of obsidian within the chipped stone assemblages at differently sized Pre-Pottery Neolithic settlements in the Levant (Ibáñez et al. 2015: Table 3). We have known for a long time that the volume of obsidian in the system increased with time across the Pre-Pottery Neolithic. And we have also known that, in addition to the increasing average size of settlements, there is an increasing range of site size in the later Pre-Pottery Neolithic. Ibáñez and his colleagues (2015) have been able to compile considerably more data for the ratio of obsidian to flint at sites in the Levant that are more than 500 km from the obsidian sources; they have also classified sites as small (<1.2 ha), medium (1.2–5.3 ha), and big (>5.3 ha). The data for the ratio of obsidian to flint at sites in relation to site size is instructive, particularly for the later Pre-Pottery Neolithic. 'Big' sites of the later Pre-Pottery Neolithic are at least six times larger in area than 'small' sites, but the 'big' sites had 33 times more obsidian than the 'small' sites that they served, and 6 times more than the 'medium' settlements. That looks like the effect of settlement scaling, whereby 'productivity' or 'wealth' increases with the scale of the settlement's population in accordance with a super-linear exponent, and these larger settlements played key roles in the regional social exchange networks.

We have aggregated populations in our Neolithic settlements, and I strongly suspect that the quantified analytical techniques of settlement scaling theory could be applied to our Neolithic aggregated settlements and networks, generating development and social change in a super-linear relation to the scale of their populations. We have generally referred to the settlements of the early Neolithic as the emergence of sedentism: should we rather see the emergence of sustainable aggregation in the Neolithic as the successful formulation of a process that continues to scale up through the Chalcolithic, to the late Chalcolithic proto-urban sites of north Mesopotamia like Tell Brak (Oates et al. 2007; Oates & Oates 1997; Ur et al. 2011), and on into the Uruk period of the later fourth and the Early Dynastic of the third millennium BCE, the cities that characterise the so-called urban revolution?

That is looking forward in time from a Neolithic origin to the early urban period and beyond towards the present, suggesting that the emergence of aggregated settlements in the Neolithic is the beginning of the settlement scaling processes that underpin contemporary urbanism worldwide. There is also a hint that we may be able to look back earlier in time in order to learn of the process that brought the first aggregated settlements into being. The most recent paper from the members of the Social Reactors Project is very significant for us in this regard. Lobo and colleagues have examined statistically the density of occupation across a large database of hunter-gatherer open sites (Lobo et al. 2022). A number of anthropologists have noted that when larger numbers of people are present at a hunter-gatherer site, the sites seem to be larger and the spacing of shelters or hearths greater. The analysis of a large cross-cultural sample of hunter-gatherer sites shows that the area occupied increases at a greater rate than is explained by the increase in numbers of people (that is, in accord with a super-linear exponent). Most hunter-gatherer groups do the opposite of what has been found with cities (and early Neolithic settlements); while cities and pre-urban settlements show population densities increase in a sub-linear relationship with settlement area, the area of hunter-gatherer sites increases at a super-linear rate relative to the numbers of people at the site.

Lobo et al. noted that the exceptions to that rule for hunter-gatherer sites was found to be among semi-sedentary or sedentary hunter-gatherer societies on the Pacific coasts of North America; these societies relied on resources like stored acorn harvests in California, seasonal salmon-runs or hunting of marine mammals in the Pacific north-west, and in general storage technologies. We have known for a long time that there is a significant difference between hunter-gatherers who live by immediate-return strategies and those that operate a delayed-return strategy (Woodburn 1982); and Alain Testart (1982) also pointed out that hunter-gatherers who harvest and store food resources are more like simple farming societies than societies dependent on mobile foraging.

As Lobo and his colleagues point out, the strategy of 'de-densification' of hunter-gatherer camps is a mechanism enabling foragers to experience the benefits of a degree of aggregation within the constraints of their social conventions, physical technologies, and economic (energetic) productivity (Lobo et al. 2022, 81). They conclude that longitudinal studies are needed to reveal the nature of the transition from mobility and 'de-densification' to aggregation and sedentism. I suggest that we have the material for such a longitudinal study in the Epipalaeolithic and Neolithic, showing how groups in southwest Asia made that diametric transition from de-densification to aggregation and agglomeration. We can produce the evidence for a move towards (seasonal) aggregations, and the early implementation of small, permanent settlements within the long millennia of the Epipalaeolithic (of the Levant, at least). And the sudden appearance in the final millennium of the Epipalaeolithic and the opening of the Pre-Pottery Neolithic of permanent settlements with substantial architecture, monumental and communal buildings, and new forms of symbolism and materialised ritual is witness to the implementation of new social norms and institutions. In the Neolithic we have the qualitative, if not yet the quantitative, evidence that documents the inception of aggregation and the revolutionary beginning of a process that links the super-linear scaling of Neolithic settlements to the phenomenon of contemporary worldwide urbanism.

The Three As: Aggregation Promotes Accelerated Cultural and Economic Accumulation

The truth is that, beginning with the ENT in southwest Asia, and soon followed by equivalent transformations in other parts of the Americas, Africa and Asia, our way of life has been founded on population growth and agricultural expansion, on aggregation and its capacity to accelerate cultural and economic innovation and cultural growth. Population growth was not something new, of course; earlier species of the genus Homo had spread out of Africa across Eurasia, and Homo sapiens had grown to populate all continents except Antarctica before the Epipalaeolithic–Neolithic with which we have been concerned. Following from the Epipalaeolithic–Neolithic transformation, however, human history has depended on the availability of more land on which to produce food to feed an ever-expanding population, and more resources such as metal ores and timber. And now we are living at the precise time when a way of life that has been foundational to our economic, social and cultural history has reached its planetary limit, and it must be changed. The exponential growth of human population may be slowing towards a plateau later this century (Dorling 2020), but not soon enough.

The ENT initiated a way of life that was predicated on the cultural, social and economic advantages of growth, and it was supported on a highly sophisticated means of ensuring cultural transmission and cultural accumulation. Cultural, social, economic and technical development has continued to expand and to accelerate until today. Accelerating tempo and scale is the characteristic of long-term human evolution, and the key feature of the ENT is the way that the cultural niche was transformed, producing a clear acceleration in the scale and tempo of human social, cultural, and economic growth. The idea of super-communities that concentrated the cultural capacity of thousands and tens of thousands of people evolved further and further until today's world is populated with cities of tens of millions of individuals and nation-states whose populations are counted in hundreds of millions of people. We have reached the climax-crisis of an evolutionary line that was first established in the ENT.

Following Gordon Childe's claim for the socio-economic success of his Neolithic revolution, it was commonly thought that the adoption of sedentary life and the development of a farming economy represented progress at last after the endless millennia of Palaeolithic, living hand to mouth by hunting and gathering. Now we can recognise that the ENT initiated the acceleration of scale and tempo that has brought us to where we are today. Acceleration of scale and tempo is very noticeable – alarming - now, even within the space of a decade, whether in the growth of global human population, growth in the consumption of natural resources, the loss of species, the removal of forests to make way for agriculture or mining, or the development of new communications apps. Can we balance continuing growth in well-being with the necessity of changing our reliance on growth and ever more demands on the Earth's resources? We have much to think about: time is not on our side, and we are challenged to move beyond the ways that our parents, grand-parents, and ancestors back as far as the ENT devised and lived their lives. We need to make a new transformation, changing our ideas of how to live well, to live together, and how to fashion economies that do not rely on the extraction and consumption of more materials.

It will help to know what made the previous transformation work. The success of the way of life that evolved in the ENT was based on the efficiency and resilience of the modes of cultural transmission; the aggregation of large numbers of people into super-communities and those modes of cultural transmission from one generation to the next ensured the security of large and complex bodies of cultural knowledge and practice, at the same time providing near-ideal conditions for innovation and cultural accumulation. We need those capacities for resilience, secure transmission of knowledge and values, cultural, social and economic innovation; but we need to be aware we must make fundamental changes to a way of life that has been 'inbred' for more than 10,000 years, dependent on expansion, intensification and growth.

Understanding the ENT as the inception of aggregation and acceleration sets the exponential expansion of our present problems into context. The development of farming at the end of the Neolithic leads to other important developments, such as the ability to accumulate and inherit wealth, the growth of competition, inequality, and warfare; but that is another story that needs telling in another place. The ENT in southwest Asia and its equivalents in other continents established the way of life that has been foundational to the rest of human history through aggregation with its acceleration of the rate of cultural–economic expansion and growth. In our time, we are inheritors of a cultural mode dependent on aggregation and accumulation and acceleration: we find ourselves at the end of a 12,000 -year-long Anthropocene, face-to-face with radical uncertainty that demands fundamental changes in the ways that most of the human population have been living for many thousand years. In our favour, we can learn from the history of the ENT that the societies from which we are descended show that we inherit a unique capacity for cultural, social and economic innovation. Can we employ those advantages of scale and the cultural and social skills for large-scale cooperation to devise and implement new ways for our societies to live together as successfully as our Neolithic ancestors; the new challenge is to scale up new solutions in a super-linear mode, while scaling down – quickly and drastically – our reliance on simply expanding and intensifying our use of energy and resources.

References

Bar-Yosef Mayer, D. E. (2013). Towards a Typology of Stone Beads in the Neolithic Levant. *Journal of Field Archaeology*, 38(2), 129–42.

Bettencourt, L., & West, G. (2010). A Unified Theory of Urban Living. *Nature*, 467(7318), 912–3. doi:10.1038/467912a

Bettencourt, L. M. A. (2013). The Origins of Scaling in Cities. *Science*, 340(6139), 1438–41. doi:10.1126/science.1235823

Bettencourt, L. M. A., Lobo, J., Helbing, D., Kühnert, C., & West, G. B. (2007). Growth, Innovation, Scaling, and the Pace of Life in Cities. *Proceedings of the National Academy of Sciences*, 104(17), 7301–6. doi:10.1073/pnas.0610172104

Bocquet-Appel, J.-P., & Bar-Yosef, O. (Eds.). (2008). *The Neolithic Demographic Transition and Its Consequences*. Dordrecht: Springer.

Crutzen, P. J. (2002). Geology of Mankind. *Nature*, 415, 23. doi:10.1038/415023a

Crutzen, P. J., & Stoermer, E. F. (2000). The "Anthropocene". *Global Change Newsletter*, 41, 14–8.

Dorling, D. (2020). *Slowdown: The End of the Great Acceleration-and Why It's Good for the Planet, the Economy, and Our Lives*. London: Yale University Press.

Dunbar, R. I. M. (2022). Managing the Stresses of Group-Living in the Transition to Village Life. *Evolutionary Human Sciences*, 1–39. doi:10.1017/ehs.2022.39

Garrard, A. N., & Byrd, B. (1992). New Dimensions to the Epipalaeolithic of the Wadi Jilat in Central Jordan. *Paléorient*, 18, 47–62.

Goren, Y., & Goring-Morris, A. N. (2008). Early Pyrotechnology in the Near East: Experimental Lime-Plaster Production at the Pre-Pottery Neolithic B Site of Kfar HaHoresh, Israel. *Geoarchaeology*, 23(6), 779–98. doi:10.1002/gea.20241

Ibáñez, J. J., Ortega, D., Campos, D., Khalidi, L., & Méndez, V. (2015). Testing Complex Networks of Interaction at the Onset of the Near Eastern Neolithic Using Modelling of

Obsidian Exchange. *Journal of The Royal Society Interface*, 12, 20150210. doi:10.1098/rsif.2015.0210

Jones, M. D., Maher, L. A., Macdonald, D. A., Ryan, C., Rambeau, C., Black, S., & Richter, T. (2016). The Environmental Setting of Epipalaeolithic Aggregation Site Kharaneh IV. *Quaternary International*, 396, 95–104. doi:10.1016/j.quaint.2015.08.092

Kuijt, I. (2000). People and Space in Early Agricultural Villages: Exploring Daily Lives, Community Size and Architecture in the Late Pre-Pottery Neolithic. *Journal of Anthropological Archaeology*, 19(1), 75–102. doi:10.1006/jaar.1999.0352

Lenton, T. M., Held, H., Kriegler, E., Hall, J. W., Lucht, W., Rahmstorf, S., & Schellnhuber, H. J. (2008). Tipping Elements in the Earth's Climate System. *Proceedings of the National Academy of Sciences*, 105(6), 1786–93. doi:10.1073/pnas.0705414105

Lobo, J., Bettencourt, L. M. A., Ortman, S. G., & Smith, M. E. (2020). Settlement Scaling Theory: Bridging the Study of Ancient and Contemporary Urban Systems. *Urban Studies*, 57, 731–47. doi:10.1177/0042098019873796

Lobo, J., Whitelaw, T., Bettencourt, L. M. A., Wiessner, P., Smith, M. E., & Ortman, S. (2022). Scaling of Hunter-Gatherer Camp Size and Human Sociality. *Current Anthropology*, 63(1), 68–94. doi:10.1086/719234

Maddin, R., Muhly, J. D., & Stech, T. (1999). Early Metalworking at Çayönü. In A. Hauptmann, E. Pernicka, T. Rehren, & Ü. Yalçin (Eds.), *The Beginnings of Metallurgy. Proceedings of the International Conference "The Beginnings of Metallurgy", Bochum 1995* (pp. 37–44). Bochum: Deutsches Bergbau Museum.

Maher, L. (2017). Late Quaternary Refugia, Aggregations, and Palaeoenvironments in the Azraq Basin, Jordan. In Y. Enzel & O. Bar-Yosef (Eds.), *Quaternary of the Levant: Environments, Climate Change, and Humans* (pp. 679–90). Cambridge: Cambridge University Press.

Maher, L., & Macdonald, D. (2013). Assessing Typo-Technological Variability in Epipalaeolithic Assemblages: Preliminary Results from Two Case Studies from the Southern Levant. In F. Borrell, M. Molist, & J. J. Ibàñez (Eds.), *The State of Stone: Terminologies, Continuities and Contexts in Near Eastern Lithics* (pp. 29–44). Berlin: Ex Oriente.

Molist, M., Montero-Ruiz, I., Clop, X., Rovira, S., Guerrero, E., & Anfruns, J. (2009). New Metallurgic Findings from the Pre-Pottery Neolithic: Tell Halula (Euphrates Valley, Syria). *Paléorient*, 35(2), 33–48.

Oates, J., McMahon, A., Karsgaard, P., Quntar, S. A., & Ur, J. (2007). Early Mesopotamian Urbanism: A New View from the North. *Antiquity*, 81(313), 585–600. doi:10.1017/S0003598X00095600

Oates, J., & Oates, D. (1997). An Open Gate: Cities of the Fourth Millennium BC (Tell Brak 1997). *Cambridge Archaeological Journal*, 7, 287–97.

Ortega, D., Ibañez, J., Khalidi, L., Méndez, V., Campos, D., & Teira, L. (2014). Towards a Multi-Agent-Based Modelling of Obsidian Exchange in the Neolithic Near East. *Journal of Archaeological Method and Theory*, 21(2), 461–85. doi:10.1007/s10816-013-9196-1

Ortman, S., & Coffey, G. (2017). Settlement Scaling in Middle-Range Societies. *American Antiquity*, 82(4), 662–82.

Ortman, S. G., Davis, K. E., Lobo, J., Smith, M. E., Bettencourt, L. M. A., & Trumbo, A. (2016). Settlement Scaling and Economic Change in the Central Andes. *Journal of Archaeological Science*, 73, 94–106. doi:10.1016/j.jas.2016.07.012

Özdogan, M., & Özdogan, A. (1999). Archaeological Evidence on the Early Metallurgy at Çayönü Tepesi. In A. Hauptmann, E. Pernicka, T. Rehren, & Ü. Yalçin (Eds.), *The Beginnings of Metallurgy. Proceedings of the International Conference "The Beginnings of Metallurgy", Bochum 1995* (pp. 13–22). Bochum: Deutsches Bergbau Museum.

Roberts, B., Thornton, C., & Pigott, V. (2009). Development of Metallurgy in Eurasia. *Antiquity*, 83(322), 1012–22. doi:10.1017/S0003598X00099312

Rockström, J., & Gaffney, O. (2021). *Breaking Boundaries: The Science behind Our Planet*. London: Dorling Kindersley.

Ruddiman, W. (2017). Geographic Evidence of the Early Anthropogenic Hypothesis. *Anthropocene*, 20, 4–14. doi:10.1016/j.ancene.2017.11.003

Ruddiman, W. F. (2010). *Plows, Plagues, and Petroleum: How Humans Took Control of Climate*. Princeton, NJ: Princeton University Press.

Ruddiman, W. F. (2013). The Anthropocene. *Annual Review of Earth and Planetary Sciences*, 41, 45–68.

Smith, B. D., & Zeder, M. A. (2013). The Onset of the Anthropocene. *Anthropocene*, 4, 8–13. doi:10.1016/j.ancene.2013.05.001

Smith, M. E. (2019). Energized Crowding and the Generative Role of Settlement Aggregation and Urbanization. In A. Gyucha (Ed.), *Coming Together: Comparative Approaches to Population Aggregation and Early Urbanization* (pp. 37–58). Albany: State University of New York Press.

Steffen, W., Rockström, J., Richardson, K., Lenton, T. M., Folke, C., Liverman, D., ... Schellnhuber, H. J. (2018). Trajectories of the Earth System in the Anthropocene. *Proceedings of the National Academy of Sciences*, 115(33), 8252–9. doi:10.1073/pnas.1810141115

Stephens, L., Fuller, D., Boivin, N., Rick, T., Gauthier, N., Kay, A., ... Ellis, E. (2019). Archaeological Assessment Reveals Earth's Early Transformation through Land Use. *Science*, 365(6456), 897–902. doi:10.1126/science.aax1192

Tarolli, P., Vanacker, V., Middelkoop, H., & Brown, A. G. (2014). Landscapes in the Anthropocene: State of the Art and Future Directions. *Anthropocene*, 6, 1–2. doi:10.1016/j.ancene.2014.11.003

Testart, A. (1982). The Significance of Food Storage among Hunter-Gatherers: Residence Patterns, Population Densities and Social Inequalities. *Current Anthropology*, 23, 523–37.

Ur, J., Karsgaard, P., & Oates, J. (2011). The Spatial Dimensions of Early Mesopotamian Urbanism: The Tell Brak Suburban Survey, 2003–2006. *Iraq*, 73, 1–19. doi:10.1017/S0021088900000061

Whitehouse, H., Mazzucato, C., Hodder, I., & Atkinson, Q. D. (2014). Modes of Religiosity and the Evolution of Social Complexity at Çatalhöyük. In I. Hodder (Ed.), *Religion at Work in a Neolithic Society: Vital Matters* (pp. 134–55). Cambridge: Cambridge University Press.

Woodburn, J. (1982). Egalitarian Societies. *Man*, 17, 431–45.

Yalçin, U., & Pernicka, E. (1999). Friühneolithische Kupfermetallurgie von Aşıklı Höyük. In H. Hauptmann, E. Pernicka, T. Rehren, & U. Yalçin (Eds.), *The Beginnings of Metallurgy. Proceedings of the Inter-National Conference "The Beginnings of Metallurgy", Bochum 1995* (pp. 45–55). Bochum: Deutsches Bergbau-Museum.

Index

Printed and bound by CPI Group (UK) Ltd, Croydon, CR0 4YY

29/09/2024

01038894-0001